The Fast Breeder Reactor

Need? Cost? Risk?

Edited by

Colin Sweet

M

First published 1980 by
THE MACMILLAN PRESS LTD
London and Basingstoke
Associated companies in Delhi Dublin
Hong Kong Johannesburg Lagos Melbourne
New York Singapore and Tokyo

Typeset by
Reproduction Drawings Ltd, Sutton, Surrey

Printed in Great Britain by
Unwin Brothers Ltd.
Old Woking
Surrey

British Library Cataloguing-in Publication Data

The fast breeder reactor.
    1. Fast reactors—Economic aspects—Congresses
    2. Fast reactors—Social aspects—Congresses
    I. Sweet, Colin
    338.4'7'6214834   HD9698.AQ

ISBN 0-333-27973-5

Contents

Appendices

iv

The Contributors

Norman Dombey, Reader in Theoretical Physics, University of Sussex

F. R. Farmer, Safety Adviser to the Atomic Energy Authority; Visiting Professor at Imperial College, London

Leslie Grainger, Formerly Chairman, IEA (Coal Services) Limited

Dr Bhupendra Jasani, Stockholm International Peace Research Institute

Dr P. M. S. Jones, Head of Economics and Programmes, Atomic Energy Authority, London

Gerald Leach, Senior Research Fellow, Institute of Environment and Development (with F. Romig, A. Van Buren and G. Foley, co-author of *Low Energy Strategies*)

Patricia J. Lindop, Professor of Radiobiology at St Bartholomew's Hospital Medical School, London (Member of the Royal Commission on Environmental Pollution)

Walter Marshall, CBE, FRS, Deputy Chairman of the Atomic Energy Authority (formerly Chief Scientist at the Department of Energy)

Peter Odell, Professor of Economic Geography, Erasmus University, Rotterdam (author of many studies on oil)

Michael J. Prior, Formerly Economist with the International Energy Agency (Coal Reseach), London; presently Senior Consultant, Environmental Resources Ltd

John Surrey, Senior Research Fellow, Science Policy Research Unit, Sussex University (with Lesley Cook, co-author of *Energy Policy: Strategies for Uncertainty*)

Colin Sweet, Senior Lecturer in Economics, Polytechnic of the South Bank, London

Peter J. Taylor, Research Student, Institute of Social Anthropology, University of Oxford; Co-director of the Political Ecology Research Group Ltd

Meredith W. Thring, Professor (and Head of Department) of Mechanical Engineering, Queen Mary College, University of London

David Widdicombe, QC, Chairman of the Administrative Law Committee of Justice

Acknowledgements

Acknowledgements are made to the Director and staff of the Polytechnic of the South Bank, London, SE1, where the conference on which this book is based was held on 23–24 November 1978.

Abbreviations
Used in the Text

ADM	atomic demolition munitions
AEA	Atomic Energy Authority (UK)
AEB	Atomic Energy Board (Canada)
AEC	Atomic Energy Commission (USA)
AFB	atmospheric fluidised bed
AGR	advanced gas-cooled reactor
BGC	British Gas Corporation
BNFL	British Nuclear Fuels Ltd
BWR	boiling water reactor
CCG	combined cycle gasification
CDFR	Commercial Demonstration Fast Reactor
CEGB	Central Electricity Generating Board
CERN	*Conseil Européen pour la Recherche Nucléaire* (European Organisation for Nuclear Research)
CFR (-1)	Commercial Fast Reactor (-1)
CHP	combined heat and power
CONAES	Committee on Nuclear and Alternative Energy Systems (USA)
EAS	Economic Assessment Service (of IEA)
EEC	European Economic Community
EPCD	Energy Policy Consultative Document
ETSU	Energy Technology Support Unit
FBC	fluidised bed combustion
FBR	fast breeder reactor
FCI	fuel–coolant interaction
FGD	flue-gas desulphurisation
GDP	gross domestic product
GNP	gross national product
HSE	Health and Safety Executive (UK)
HT(G)R	high-temperature (gas-cooled) reactor
HWR	heavy-water reactor

IAEA	International Atomic Energy Agency
ICRP	International Commission on Radiological Protection
IEA	International Energy Agency
IIASA	International Institute of Applied Systems Analysis
INFCE	International Nuclear Fuel Cycle Evaluation
JET	Joint European Torus
LMFBR	liquid-metal fast breeder reactor
LWR	light-water reactor
MRC	Medical Research Council
NCB	National Coal Board
NII	Nuclear Installations Inspectorate (UK)
NNC	National Nuclear Corporation
NPT	Non-Proliferation Treaty (of nuclear weapons)
NRC	Nuclear Regulatory Commission (USA)
NRPB	National Radiological Protection Board
OECD	Organisation for Economic Cooperation and Development
OPEC	Organisation of Petroleum-Exporting Countries
PERG	Political Ecology Research Group
PF	pulverised fuel
PFB	pressurised fluidised bed
PFR	Prototype Fast Reactor
PNE	peaceful use of nuclear explosives
PWR	pressurised water reactor
RCEP	Royal Commission on Environmental Pollution
R & D	research and development
SALT	Strategic Arms Limitation Treaty
SNG	synthetic natural gas
SSEB	South of Scotland Electricity Board
THORP	Thermal Oxide Reprocessing Plant
UKAEA	United Kingdom Atomic Energy Authority
UNSCEAR	United Nations Scientific Committee on the Effects of Atomic Radiation

Introduction

Colin Sweet

The UK has yet to decide if it is going to proceed to the construction of a programme of commercial fast breeder reactors. The decision may well be taken in the lifetime of the present Government. Whichever way it goes, it will have a major impact not only domestically but internationally. Relations with the USA and with the EEC are deeply involved with nuclear strategy. The nuclear fuel cycle is a world-wide industry, with fuel supplies located in only a few countries, while enrichment and reprocessing facilities are even more restricted in their availability. Domestically, a decision to proceed will mean commanding more resources than for any other technology. The political and social implications may be even greater. A decision not to proceed will also have its impact because it may mean the gradual dismantling of the nuclear industry with its many agencies and suppliers. Constructively, it will call for a radical reassessment of our energy future and more research on renewable energy sources.

Because such large resources have been committed to the fast breeder over the last 25 years, there must be a presumption that the decision will be in favour of proceeding. A great deal depends, however, on how the decision is taken, by what political processes, and within what framework of reference. This, rather than the ability of the industry to build and operate fast reactors, or the capacity of our economy to find the resources (important as these questions are), will be the most difficult and yet the most crucial element to be resolved. That is to say that the decision cannot be separated from the nature of the decision-making process.

The Sixth Report of the Royal Commission on Environmental Pollution (RCEP) provided an opening to the specifics of the fast reactor discussion that was, for its time, a model. This was not sustained by the Parker Report on Windscale, over which there was widespread and justifiable dissatisfaction; not only or even with its conclusions as with the way in which they were reached. The resultant of these two events has been the proposal by the previous Minister for the Environment, Peter Shore, that a Public Inquiry Commission should be set up to review the wider issues of the fast breeder and the quasi-legal form of the Windscale Inquiry used only for hearing the arguments on the siting of a fast reactor. In a sense, both the Sixth Report of the RCEP and the Parker Report are part of the past. In the short space of three years, social awareness has increased so rapidly that the idea of marrying a Public Inquiry Commission to a quasi-legal inquiry under a judge, which is a step forward perhaps, is not in itself adequate. Institutionally, it does not take account of the new level of social awareness. The difference between the immediate past and the immediate future can be instanced under two headings.

Firstly, the need for a great deal more information than we already possess. To illustrate the point, the fact that Sir John Hill, who is Chairman of both the UK Atomic Energy Authority (AEA) and British Nuclear Fuels Ltd (BNFL), has revealed that there was a partial core melt at Chapel Cross 15 years ago. A serious accident was averted, but to publicise this after the accident at Three Mile Island in order to avert concern illustrates how perilous is the relationship between the nuclear industry and the public.

Mr Benn, when he was the Minister, inaugurated the Open Debate on Energy, and no doubt he tried hard to fulfil his promise. But it can be argued with some justification that the Open Debate has hardly begun. Where, for example, are the reports of the accidents at Hinkley Point 'B' and Hunterston 'B' in 1977? In the files of the Central Electricity Generating Board (CEGB) and the Nuclear Installation Inspectorate, to which only a few people have access. Or, how much do we really know about the cost of nuclear power? Despite the constant repetition that it is the cheap option, the former Minister for Energy undermined this in February 1979, not by saying energy was dear, but by admitting that he *didn't know* what the cost was. 'Only a year ago', Mr Benn was reported as saying (in the *Financial Times* of 2 February 1979), 'when we were having a great argument about systems, I was being asked by some interests to commit £24 billion of public money to a pressurised water reactor system (PWR) . . . . I didn't want to commit that amount of money because *the true costs of nuclear haven't even been identified*' [my italics] . On that occasion, the decision went in favour of the advanced gas-cooled reactor (AGR) system with two more reactors to be built and followed most likely by the first PWR. This was a decision in which public acceptability was not allowed to count for anything, because no information was given to the public. It might very well have been the case that, if the full account of the AGR history had been made available, a different decision would have been taken. One authority on the subject has bracketed the AGR with Concorde as the two most costly errors in history[1]. The programme of five AGR reactors has accumulated 28 construction years overrun and its present losses cannot be less than £3000 million. It ought not to be repeated without some inquiry as to what went wrong. But, as Professor Henderson then commented[2], not only were the losses not disclosed, and not only is there no provision for collecting together and publishing the annual costs, but 'it is evident that in the case of the AGR not only need this not occur, but that virtually no one is disturbed by it.'

Secondly, the methodology or the framework of reference that is adopted in arriving at a decision will affect very decisively the nature of the decision itself. It is not only possible, but indeed likely, that, unless the discussion is sufficiently prolonged and informed, the criteria given for the decision may be irrelevant to the subject itself. It is not difficult to find such examples in the past. This does not mean that the public interest is subverted by malevolent administrators or self-seeking corporate interests, though the existence of both is not to be denied. The reasons are far more complex and they are to do with the nature of our decision-making process as it now is. Different levels of discussion are projected using different criteria, and designed to serve different functions. If there is to be a public inquiry before a decision is taken on the fast breeder, it will have to be thorough enough and provoking enough in order that these different levels of discussion can interpenetrate. The end result ought to be a criterion, or a set of criteria, from which it is understood the decision, whatever it may be, derives its meaning and relevance.

In the case of the fast breeder, the methodology will have some novel features. This will be an exceptional decision. I give three reasons for saying this. Firstly, there is the time factor. The fast breeder was postulated by scientists, both in Britain and America, in the very early days of the nuclear programme, as the objective to which they should move as rapidly as possible. At that time, the only constraints that were envisaged by those in the industry were ones of engineering feasibility. As a result, experimental fast breeder reactors were built at an early stage and well before any commercial thermal reactors were working. That was in the 1950s. At the earliest, it will be 1990 before a commercial demonstration fast reactor is operating in the UK, and several years after that before a commercial reactor programme could begin. It will not be before the second decade of the next century, at the earliest, that a fast reactor programme of any significance could be working in this country. It could be another two decades after that before fast reactors are generating most of our electricity. This makes the time span from conception to effective presence around 100 years. A decision in 1982 to proceed would then be a decision to have a complex of fast reactors and the fuel cycle factories dotted over Britain in 50 years time. How do we take decisions over such a time span? Moreover, this is not just a matter of the social or moral responsibilities involved. The time factor also dominates the economic and engineering logistics of the fast reactor because of the peculiar problem of supply and demand of plutonium and uranium fuels. The problem of time is therefore one to which we have to begin to address ourselves in a serious way. At present, we have no way of deciding what are the right decisions for such long time periods. Secondly, the decision to proceed with the fast reactor programme will be done in the face of the existence of unknown, and possibly unknowable, features. This is one area where social awareness has not only made itself felt, but where it affects the methodology. Hitherto, the unknown technological factors have been treated as matters for technical experts. There remains a strong body of opinion that says that the public's right to decide is qualified by the fact that on technical matters it is the expert who must decide. This view has been challenged by that which says technical matters are of social concern and, therefore, socially determined. For example, the satisfactory disposal of waste or the satisfactory capture of isotopic gases from a nuclear plant is something upon which the community must decide. Values and belief systems of quite a complex kind are involved and, until a consensus is arrived at, the right technical solutions cannot be properly identified. The third factor that makes fast breeder decisions exceptional is that in certain respects they are, or they may become, irreversible. The fission process is dependent upon the existence of the isotope ^{235}U which exists in very small quantities in the Earth's crust. In order to derive from this small quantity of material, energy of very large magnitudes, it is necessary to proceed along a path of increasingly high technology. As the complexities of the inter-related process become greater, we

3

may approach a point where the total process cannot be reversed. At that point, we would be very heavily dependent on fast reactor systems. The rising output of fission products would be irreversible. If, therefore, a watershed stage should arrive when the dependence on a fast reactor system would be difficult to change, we have to give consideration to that now. Alvin Weinberg's famous description of nuclear power as a Faustian bargain between the scientist and society expresses the potential development in a suitably dramatic way.

The theoretical problem that is thrown up concerns the acceptable risk to a community. The treatment of the subject of risk by Lord Rothschild[3] was meant to reassure us on this point. However, we should have few, if any, doubts that not only do we have no way of answering the question of acceptable risks, but for the present at least we must endure the most spurious treatment of the subject undertaken in order to rationalise the unknowable risks which nuclear technology currently involves.

The logistical requirements of a plutonium economy have to be measured on the social plane as well. This has yet to be fully explored. What is involved is our capacity to learn from the past in order to control the future. But who is to control and who is to be controlled. Perception in these matters is growing and the more it does so, the more it poses the challenging question as to whether we have a methodology that can handle decision making for the fast breeder technology. The RCEP advised that one should not proceed with the fast breeder until there was no reasonable doubt that it was beneficial. The Energy Green Paper rejects that advice and places the fast breeder as the number-one decision to be taken. How much time do we have before a decision has to be made? Perhaps this is the first matter that a Public Inquiry Commission should explore.

Creating a context in which the fast reactor decision can be debated is therefore no easy matter. The problem will be to establish dimensions of sufficient breadth and depth as to encompass the economic, political, social and technological questions that are relevant to the exceptional nature of the subject. This volume makes a limited contribution. We are fortunate to have the views of three senior members of the UK Atomic Energy Authority and they are important for that reason alone. The remainder of the papers are heterogeneous in their treatment of the subject. It is necessary to subject the official view to a critical assessment and it was the purpose of the conference at the Polytechnic of the South Bank to do that. But the issues that need clarifying go far beyond what is to be found in the official literature. The reader will perceive these for him or herself. There is no question of trying in this volume to strike 'a balanced view,' because there is no way in which such a phrase can be interpreted. We interpret our task at this stage to expose all aspects of the problem to a questioning analysis in order that there may be eventual clarity on what the important criteria are.

The first five papers seek to put the fast breeder in the context of existing energy policy. This may be a limited exercise, because it is one that largely has to accept the parameters of official policy. These are almost totally economic and make no significant reference to the political or international context. (Implicity, they do raise important social matters.) Yet this is an exercise which is very necessary. Decision making about the fast breeder certainly should not be made solely on the energy supply-and-demand perspectives. But it is equally true that there can be no rational debate without such perspectives. An energy scenario is a necessary condition in energy decision making. The various critical appraisals which are to be found in the first section are not unrelated to the reconsideration of energy forecasting that can be observed to be taking place at various levels today. The question of methodology is raised most especially in Leach's paper. The output of the study teams sponsored both in the USA and in the UK[4] by the Ford Foundation and represented in this volume by Gerald Leach's paper has yet to be fully assessed. But because it does present a quite different methodology and follows it through to an impressive range of conclusions, it is an important point of departure. The method used by Leach and his colleagues is simple to understand and it offers a direct alternative to that of the official scenario. Its objectives are defined in end-use energy terms, i.e. social needs, and it derives its primary energy input from this objective via a highly detailed sectoral analysis. It is a physical energy output–input model and its merit lies in its emphasis on efficiency. This is a relatively high-technology scenario not to be confused with the alternative energy scenarios that proceed from a different view of the supply side.

The section on economics (Part III) also raises its methodological points, though not very explicitly. But they are important ones. Firstly, it is readily apparent to the reader that we have, as yet, no proper way of assessing the costs of a fast breeder programme. Therefore, there is no meaning, as yet, to the claim that the fast breeder brings economic benefit. *No one* knows if such claims are true or false. The time horizon is far too long for any meaningful projection of economic variables. The decision, should it be taken and on economic grounds, would be taken using data that pretend to have a precision that they do not possess. We could usefully begin by giving some reasonably accurate answers about past costs and in so doing we would clear up some of the controversial and unsatisfactory features of the present method of economic assessment and accounting. In Energy Working Paper No. 21[5], the Department of Energy introduced the qualification to its figures on nuclear costs, which states that the published costs of nuclear power are no guide to future choice in power stations. It remains an open question whether the fast reactor will be economical, but the first problem is to decide if there has to be an economic test that the breeder has to pass and how that test is to be devised. The admission by the AEA that the first large reactor of 1300 MW will be constructed at a loss is itself something novel in the electricity supply industry. It is for this reason, presumably, that the reactor, which was originally labelled CFR-1, has not been designated CDFR (commercial demonstration fast reactor). There would be some objections to building a commercial reactor which was known in advance to be a financial loser.

The announcement by the CEGB recently (March 1979) that the cost of AGR output, at current prices, for reactors to be commissioned in 1985 will be 1.8p/kWh also represents a significant change. This was at a time when the AEA was publishing a new and unrealistic price of 1.23p/kWh. Yet this in itself was a considerable sum considering that the last published price for nuclear power by the CEGB was 0.78p/kWh. The most recent figure of 1.8p is, therefore, doubly welcome because (a) it brings stated costs to a more realistic level and ends the myth that nuclear power is cheap, and (b) it suggests that the CEGB is differentiating its role from that of the AEA and BNFL. As a buyer of technology and materials from AEA/ BNFL, and as a seller of power to the consumer, the CEGB should logically be concerned to pursue a different role. This is important if the CEGB as the buyer of the commercial fast breeder reactors is to apply some economic tests and safeguard against the misuse of resources. The monolithicism that has so characterised the three agencies since the decision was taken in 1965 to launch the AGR programme has had results which can only be described as lamentable. The CEGB, as the greatest capital spender in our economy, should show a scrupulous regard for economic welfare. It is as a critic rather than a patron of the supplying agencies in both the public and the private sector that demonstrates that the CEGB is fulfilling such a role.

An awareness of the inadequacy of the repeated claim that nuclear power is the cheapest option may be responsible for the emergence in the official literature of the use of cost–benefit analysis. At the Windscale Inquiry, the submission made by BNFL contained the claim, since repeated, that reprocessing of nuclear waste is a benefit on two grounds. Firstly, it is a conservation measure. It has been recently asserted that if all the 20 000 tonnes of nuclear waste in the UK were utilised in fast breeders, this would meet all our energy needs for 260 years[6]. (This of course, is a purely theoretical statement although it was not intended as such.) The second consideration is that reprocessing and recycling into fast reactors is socially beneficial because it is a method of waste disposal and it reduces, rather than increases, the amount of highly radioactive isotopes (notably plutonium) in the environment. Whether reprocessing can be said to close the fuel cycle is a matter of debate. Dr Walter Marshall's thesis on the fast breeder as an incinerator rather than a breeder of plutonium is now well known and to the environmental benefits he adds a third—namely, that by recycling plutonium and reprocessing and refabricating in a way that retains some of the highly active fission products and actinides, the dangers of weapon proliferation via the 'plutonium economy' can be appreciably reduced, if not eliminated.

The introduction of relevant issues, such as waste disposal, decommissioning, proliferation, civil liberties, etc, which have previously been neglected, is to be welcomed. But the suggestion that they can be brought into some decision-making framework by the use of cost–benefit analysis has only underlined how far we are from establishing a credible framework. Cost–benefit analysis is a method which is open to considerable abuse, and one authoritative work[7] in this area has argued very cogently that it is precisely the issues of proliferation, civil liberties and radiation hazards that 'make social cost–benefit analysis largely, though not entirely, irrelevant to the evaluation of energy futures.' They add further that 'it is very arguable that cost–benefit analysis has had little significant success in environmental analysis.' The essential reasons for this lie in three areas: (a) the conceptual basis of cost-benefit analysis, (b) the use of discount rates makes long-term costs (e.g. waste disposal) appear totally insignificant, and (c) the difficulty, bordering upon impossibility, of placing quantitative values on important factors in the nuclear balance sheet.

The issue of cost–benefit analysis is not entirely irrelevant (as Professor Pearce argues), because it brings to the surface the question of whether particular aspects of the nuclear fuel cycle should be assessed for economic benefits or for social costs. The building of the THORP reprocessing plant is a case in point. If it is socially necessary, then it is a cost and not an economic benefit. If it is a benefit, however, then it has to be shown that building the plant is a proper use of resources, i.e. it will bring an average return on the capital invested. If the first fast breeder is not to be assessed on economic grounds, but taken to be an R & D project, then John Surrey's point is valid that the opportunity cost of the fast breeder must be looked at in terms of the benefits foregone on alternative energy research. This was not understood by the nuclear industry at the Windscale Inquiry and is a point still to be argued. The notion that nuclear power will bring benefits because spent fuels can be used in fast breeder programmes means that, by benefits, we are talking about conservation of energy resources and possibly waste disposal as well (although the latter point is debatable). These are not necessarily benefits in the economic sense and, if they were, they would in any case be realised by the producers independently. If the case for reprocessing and progressing towards the 'plutonium economy' is conservation of energy resources, then we are talking about something quite different. It would mean that we would be starting from a quite different premise and that energy policy as a whole would then have to be reconstructed. If this is not the case, however, and we are to follow conventional guidelines for the public sector in a market economy, then the nuclear industry has to demonstrate that because of conservation needs it is a special case; that it may be permissible to raise real costs for social reasons or that it should use a zero discount rate. What is not permissible is to argue both cases at the same time.

While the task of making economic assessments and projecting energy scenarios has many problems, it is in Parts II and IV of this volume that the more intractable problems arise. As the present state of the discussion so clearly shows, in the fields of risk, safety, proliferation, civil liberties, etc, progress is slow. What is potentially serious for the energy debate is that official energy policy has progressed most slowly of all. In fact, it is being out-distanced to a degree that suggests that, if it is ever to catch up, then there will have to be some major change of outlook in the Department of Energy—which on the face of it seems unlikely! The Energy Green Paper offers views only in segments on energy policy, i.e. energy scenarios and economic assessment. (Even at that level, it is not very effective because it is apparent that it arrives at its projection by holding court to all the energy industries, with the result that the overall forecasts are too optimistic and disaggregated that they become mutually inconsistent.) Indeed, if we compare the Green Paper with the Fuel Policy White Paper of 1967, what is striking is how little has changed in 12 years in the manner and content of policy formulation. In 1967, the basic philosophy was to let the market forces dispose of energy supply. The theoretical basis of the scenarios was marginal cost resource analysis which, as an instrument to guide energy dispartition and strategy, suffers from severe difficulties in its practical application. The criteria have been re-ordered but only in minor ways despite the great changes of the last 12 years. Long-term security of supply is, if anything, more firmly established as the priority in the formulation of energy strategy. But, even here, the interpretation of such a priority lacks a great deal if one looks for its rational interpretation in the energy mix, rates of resource depletion, and treatment of energy efficiencies in the supply side favoured by the Green Paper. The same is true for efficient utilisation of both primary energy and end-use energy. The list could be lengthened.

If energy needs are such that security of supply is the objective, and hence the basis of policy and planning in the energy industries, then we have to rework the

7

various parts of the energy supply industry into such a framework. What seems to me quite inadmissible is that social needs should be appealed to as criteria when it is convenient to do so, but otherwise disregarded. Such a tendency is now manifest where appeals are made, especially in the context of nuclear power programmes of a nature which appear to maximise the benefits of the programme on all possible counts, but which are mutually inconsistent in themselves and are not the basis of energy policy overall.

It is readily apparent therefore that, while long-term security of supply is acknowledged as the central requirement of state energy policies, the translation of this objective into a strategy is no easy task. It is further evident that, even if the resolution of the constituent parts of the energy scenario is successfully achieved, the exercise is nevertheless one that does not span the energy problems in our society. It is an exercise largely confined to the economic variables and excludes non-economic factors. This constitutes a very real problem. The treatment of risk, safety, proliferation and civil liberties as 'fringe' issues in the energy debate may accord with the political structure to which the establishment belief would confine the debate. But in the long run this will not do: it vitiates the meaning of the debate.

The stage at which we have now arrived, and which will make the context of the Fast Breeder Inquiry different from that of the Windscale Inquiry, is one in which the resolution of the conflicting issues depends essentially on the political determinants. What do we mean by political determinants? We attempt to answer that question by distinguishing between the political and economic determinants in the matrix (Appendix 4). The list is a long one and it is meant to be as inclusive as possible. There is no attempt to rank the issues in their importance, or give any quantitative or qualitative expression to them. The purpose is to list all of the issues, political, economic and social, and to relate them together, not in terms of their importance according to a set of value judgements, but to relate them together in a casual way. (The set of casual relationships represents the author's interpretation of reality, and it is not expected that everyone will agree with them.)

The difference between a methodology that takes such a matrix as a starting point and that of official energy scenarios is, of course, very large. To some, they will be seen as worlds apart, and in certain respects they are. But, ten years ago, energy was not a debatable issue; today, it is a foremost area of change. Energy practitioners (as a reading of the consultative documents that Whitehall periodically publish) imply the belief that they can continue to 'write in' the economic parameters large and bold, and to leave political determinants to the penumbra of occasional ministerial comment. For the purposes of energy scenario making therefore, the political factors are assumed, and this is convenient because the assumptions are the ones that are held by the decision makers.

This is not intended to be an attack upon the impartiality of Whitehall. It is assumed that no intelligent person believes in such pieces of constitutional ambience. My purpose is more serious, and it is to ask the question—how can we resolve such problems as the fast breeder, which take us to the frontier of our ability to take political decisions about the future, if the aspects of the problem which are clearly present in public opinion are defined out of existence by the structure of the official debate? Policy makers, as they have done for so long, use their methodology to produce conclusions that act as barriers against questions

8

they either don't understand, or don't want to discuss. It is an illustration of the unsatisfactory level of communication in the debate that the Green Paper lists six questions upon which decisions are required in the next two years. They are in the order given:

(1) The fast reactor.
(2) Other power stations to meet electricity demand growth.
(3) The building of a gas pipeline.
(4) Reinforcement of energy conservation.
(5) An energy R & D strategy that establishes priorities (no indication of what there are, because it is known that the fast breeder will take almost all of the funds!)
(6) How to make energy prices reflect real energy costs?—a sublime quest indeed.

Are these really the issues which the energy debate should resolve? One of the more interesting results of separating the economic from the political determinants is that, by almost any reckoning, the political determinants have the weight on their side. The weight of decision making is political, while the language of decision making eschews the political. This flaw in the methodology of the official scenarios has its own logic. Its practical adequacy has been its principal justification. It is not likely, however, that the decision on the fast breeder can be taken only in economic terms. The recognition of the need for a new methodology may become one of the unexpected benefits of a public inquiry.

REFERENCES

1. Henderson, P. D., 'Two Costly British Errors', *Oxford Economic Papers*, Inaugural lecture at University College, London, 1976
2. Henderson, P. D., 'The Unimportance of Being Right', *The Listener,* November 1977
3. Lord Rothschild, Dimbleby Lecture, *The Listener*, 23 November 1978
4. Leach, G., Romig, F., van Buren, A. and Foley, G., Low Energy Strategies, *Science Reviews*, 1979
5. Department of Energy, Working Paper No. 21, *Energy Strategies*
6. Moore, J. (Director, Fast Reactor Systems, AEA), Institution of Nuclear Engineers, London, 27 March 1979, *Atom*, May 1979, p. 114
7. Pearce, D. W., Edwards L. and Beuret, G., *Decision Making for Energy Futures*, The Macmillan Press, London and Basingstoke, 1979

Part I
The Fast Breeder in Energy Policy

1

Energy Perspectives for the UK

Leslie Grainger

1.1 INTRODUCTION

Energy policy has become an important feature of government in the UK. A consultative document on Energy Policy (cmnd 7101) was issued in February 1978. This was the first such document for a long time—in fact, since the discussions on energy policy in 1966 and 1967, the outcome of which seemed to reduce the enthusiasm of government for energy policy statements. An Energy Commission has now been set up, papers have been prepared and made public, including the minutes of the Commission's meetings. Even prior to this, and coincidental with the Energy Conference set up by Mr Tony Benn in 1976, was a document on R & D strategy (Energy Paper No. 11). Annual reports on the R & D programmes are now published regularly. We have had the Windscale Inquiry and a statement by Mr Peter Shore that similar Inquiries, possibly with some features reflecting the experience of the Windscale Inquiry, but reviewing particular proposals in relation to the whole, will be a regular part of the development of national energy policy. Thus, there is now a great deal more information than for some years and it is therefore appropriate that a paper on energy perspectives should centre on the Energy Policy Consultative Document (EPCD).

1.2 CURRENT STATE OF UK GRAND SUPPLY PROJECTIONS

Figure 1.1 shows the general conclusions of EPCD*. This graph is based on the 'higher' growth rate of those that were studied; the growth rates used are primarily economic growth rates. The figure shows that, by the end of the century, there is both a 'policy gap' and an allowance for new imports, the two totalling 95 million tons of coal equivalent (mtce). This is shown also in *Table 1.1*. However, in *Table 1.2* we can see the extent to which energy demand may be affected by two different (but not very different) growth rates. It will be noted that the lower growth rate would reduce demand sufficiently to bring about a small surplus (a turn round of about 100 tce) in comparison with the demand shown in *Figure 1.1*

*See Appendix 1 for official text from EPCD.

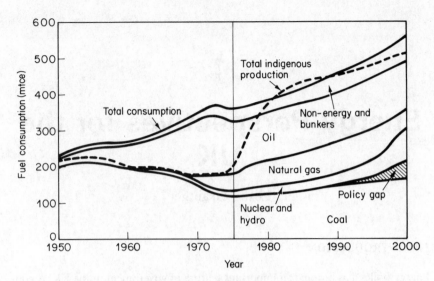

Figure 1.1 UK primary fuel consumption.

TABLE 1.1 ENERGY BALANCE (million tons of coal equivalent—mtce)

		1975	1985	2000
Demand				
Energy		314	375	490
Non-energy and bunkers		27	40	70
Primary fuel demand		341	415	560
Supply				
Indigenous:	coal	126	135	170
	natural gas	53	65	50
	nuclear and hydro	13	25	95
	oil	3	210	150
Total		195	435	465
Policy gap		0	0	50
Net imports		146	−20	45

and *Table 1.1*. It should also be stressed that the lower growth is entirely in economic growth terms and no assumptions are made about different rates of gearing between economic growth and energy demand. With various permutations, the results for potential demand in the year 2000 could cover a very wide range and for later years the range would be even wider.

Referring again to *Figure 1.1*, perhaps the most striking feature is the very rapid growth in indigenous production. This reflects, of course, the rapidly increasing output of oil from the North Sea. The Figure also shows a rapid decline in natural gas in the late 1990s matched at the same time by a rapid growth in nuclear power. The impression has more recently been given that British Gas

14

TABLE 1.2 UK ENERGY DEMAND 1975, 1985, 2000 (mtce)

	1975	*1985*	*2000*
Higher growth			
Energy	314	375	490
Non-energy and bunkers	27	40	70
Primary fuel demand	341	415	560
Lower growth			
Energy	314	350	390
Non-energy and bunkers	27	40	60
Primary fuel demand	341	390	450

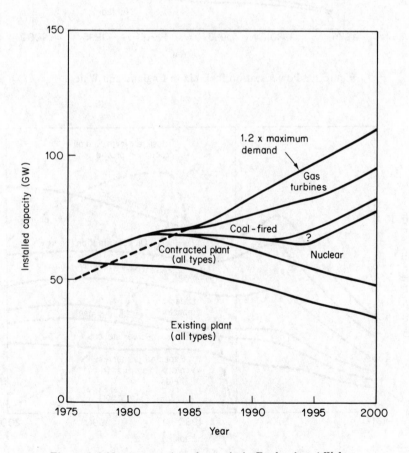

Figure 1.2 New generating plant mix in England and Wales.

15

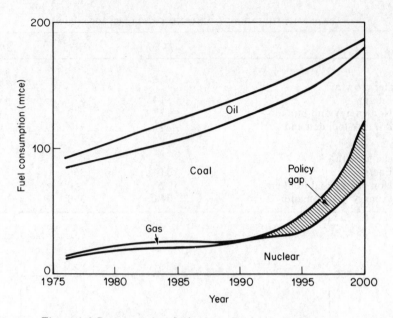

Figure 1.3 Power station fuel mix in England and Wales.

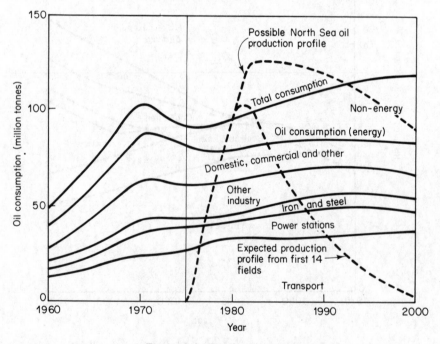

Figure 1.4 UK oil consumption.

16

would probably not now support this assumed rapid decline in natural gas supplies at that time and this, in turn, must have direct relevance to the rapid growth in nuclear power indicated. *Figure 1.2* shows the new generating plant mix in England and Wales taken from EPCD. The figure shows a rapid growth in electricity generating capacity, by a factor of more than two between now and the end of the century. It appears also that a policy of having 20% surplus capacity is maintained throughout the period. The effect of rapid growth in nuclear capacity in the second half of the 1990s is shown in *Figure 1.3*. The fuel mix indicates a very rapid decline in coal burning. Another interesting feature of *Figure 1.2* is the relatively large gas turbine capacity which is assumed to expand rapidly after 1985. This is probably a factor in the substantial use of oil in power stations throughout most of the next couple of decades; this feature is shown in *Figure 1.4*.

1.3 INTERNATIONAL AND LONGER-TERM SCENE

It is not possible to consider the UK energy position by itself. Furthermore, although the picture to the end of the century is interesting, the slopes of certain lines on the graphs exhibited make it obvious that one needs to see the following 25–50 years to develop a real policy. Both these points are illustrated by reference to the World Energy Conference conservation studies (for which Richard Eden,

TABLE 1.3 POTENTIAL WORLD PRIMARY ENERGY DEMAND (exajoules–EJ)

	1975	*2020*
World total consumption in 1975 including 26 EJ wood fuel	276	
Unmodified historical trends (at constant growth rates)		
1960–75 trend continued		1952
1933–75 trend continued		1770
1925–75 trend continued		1242
Projections found from GNP assumptions		
Low economic growth (3.0% per annum):		
L1. No price response		1128
L2. Medium price response		953
L3. High price response		897
L4. High price response plus oil constraints		840
High economic growth (4.2% per annum):		
H1. No price response		1831
H2. Medium price response		1564
H3. High price response		1455
H4. High price response plus oil constraints		1322
H5. High price response plus energy constraints		990
Potential world energy supply in 2020		820–1010

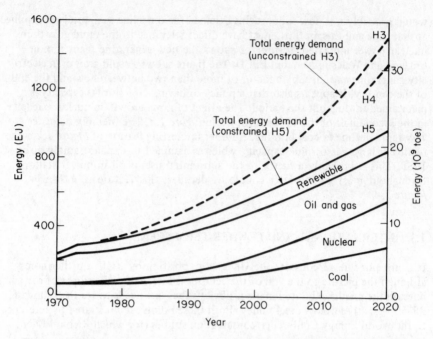

Figure 1.5 World energy demand and supply (in exajoules and in tons of oil equivalent), assuming high economic growth.

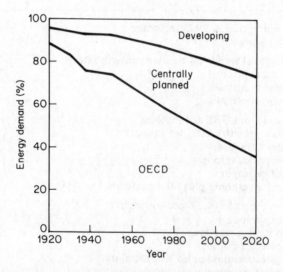

Figure 1.6 World energy demand, assuming unconstrained high growth H3 (see *Table 1.3*). Percentage shares (excluding wood) are given.

Cambridge University, was Consultant). This was a very far-reaching and complicated study, which cannot be dealt with fully here, and *Table 1.3* and *Figures 1.5–1.7* are chosen to illustrate a few particular points. First of all, the energy shortage may hit the world much sooner than is projected for the UK in EPCD. This possibility is well known, but it is important to realise how substantially and rapidly the gap widens after the year 2000. It is suggested that the OECD share will drop very significantly—some people might say this is a fall from an unfair level to a level of fair shares. However, a reduction in the share consumed does seem inevitable and, if the supply is less than forecast, it seems obvious that the OECD countries will have to feel the pinch.

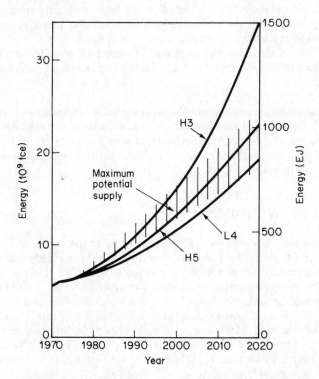

Figure 1.7 World energy demand and potential supply (in exajoules and in tons of coal equivalent).

1.4 GENERAL POINTS

Whatever one may feel about the validity of the central view presented in the EPCD, it has to be acknowledged that the assumptions are reasonably spelled out and the numbers are presented in such a way that alternative views can readily be derived. Perhaps, in fact, a wider range of possibilities should be examined and

19

presented in future documents and certainly these studies should be continued over a substantially longer period. This is especially true for technologies such as nuclear power and, particularly, for the fast breeder reactor. The starting point for the further development of energy policy, however, must be a consideration of the objectives for such a study. Mr Benn's proposals were spelled out in para. 2.2 of the document, and they remain important as objectives today:

(i) everyone can afford adequate heat and light at home;
(ii) industry's needs for energy are fulfilled at a price which reflects full resource cost and has regard to the long-term availability of the various fuels;
(iii) these objectives are met on a long-term basis, taking account of risks; the depletion of our reserves of oil and gas is regulated; research and development in energy supply and use is adequately funded; and investment in energy industries to meet these objectives is properly planned;
(iv) freedom of the consumer to choose between fuels provided at a minimum price which reflect economic cost should, where possible, be maintained and increased.

These objectives are unobjectionable as statements of intent, but it should be appreciated that all of them have very important economic implications. The quantitative working out of these economics does not seem yet to have been carried out very thoroughly and this is indeed a difficult task, especially since there are some in-built conflicts.

1.5 SOME SPECIFIC POINTS

The future demand is shown in EPCD as continuing to rise regularly. Obviously, the demand must saturate at some point in time, but this is clearly regarded as outside the time scale of the study. Price increases will affect the date of saturation, as will conservation measures in their own right. However, it does appear that the traditional understanding of what 'conservation' is about is too limited; it seems presently to be devoted mainly to avoiding the loss of heat from buildings and things of that kind. Much more important may be (a) improvements in conversion efficiency, (b) the optimum matching of primary fuel supply to end uses including non-energy uses and (c) efficiency of distribution, reserve capacity and storage (or its converse, provision for peak loading).

The document still seems too compartmentalised and therefore to reflect rather too much the aspirations of the individual energy industries rather than to formulate a separate objective judgement.

There are, for example, a number of questions to be asked with regard to the electricity growth rate which is presented. Is this extra electricity supply the best way of meeting the various needs of consumers? How well does electricity supply contribute to the flexibility required to meet the actual growth rates compared with the predictions when capital equipment is ordered, for energy and for electricity? Will the extra capacity make the best use of the primary fuel, especially in times of increasing costs? To what extent should electricity growth be linked to combined heat and power?

Within the electricity sector, the nuclear power proportion is particularly interesting. This involves high capital costs and large units, which together could reduce flexibility if there is too rapid an increase in the proportion of nuclear power in the system. The one-time excessive zeal for an all-electric, all-nuclear energy system has now been tempered, but it remains to be asked whether this process has gone far enough.

None the less, nuclear power and coal, together with conservation, are the mainstays of the future supply position in the UK. If the demand forecasts are anywhere near right, both are going to need a very great deal of effort, investment and good management to succeed in meeting their objectives around the turn of the century.

1.6 CONCLUDING THOUGHTS

The UK is remarkably well placed, through indigenous resources and acquired technology, in both the short to medium term, to meet the country's energy demand; in fact, for these time scales the problem is more of the nature of handling surpluses without destroying future flexibility and capacity, as well as morale. It would be possible to pass from glut to famine without experiencing any period of relative stability at all.

I am inclined to believe that the turnover point from adequate supplies to shortages could be rather later than suggested in EPCD. Hopefully, this may result from higher efficiency in use, including well known conservation measures, and also from better supplies of indigenous gas and oil than indicated. Although we should be prepared for the maximum rate of growth of energy demand, the danger of over-commitment should be avoided. This applies especially to electricity and, particularly, to the nuclear power sectors in the second half of the 1990s. To be prepared requires two things—a better and more open analysis of energy policy and enough R & D of the right kind. A start, belatedly, has been made on both of these, but there is a long way to go. Analytical methods are still too blunt, the R & D programme is still unbalanced with nuclear power taking too high a proportion. A lot of non-nuclear energy R & D funding is on projects in the exploratory phase. There is a need to ensure that promising aspects from this phase are given full backing in the next phases, which are bound to be very much more expensive.

In the analytical studies, there is a need to explore more deeply what sort of United Kingdom we want to have in the longer term. The link between energy usage and prosperity is frequently mentioned. Perhaps, however, it is even more important to link energy supply in the UK to the direction and location of industrial development, especially since our medium-term supply position gives us a great advantage. We shall then be able to consider perspectives more comprehensively and more accurately.

2

Some Policy Aspects of the Fast Reactor Question

John Surrey

It is now better understood that energy policy involves planning under major uncertainties rather than optimisation based on forecasts of demand and costs of supply. The key decisions are long-term ones and they must not be dominated by short-term considerations. Long-term strategic guidelines are therefore needed to provide a framework for medium-term planning by the fuel industries and for related infrastructure decisions, e.g. transport and urban redevelopment, which will influence energy demand in the long term[1]. As time horizons have lengthened, the policy emphasis has shifted to seeking insurance against undesirable outcomes and retaining and developing options. A major contingency to be guarded against is a swift return to heavy dependence on oil imports at very high costs in terms of the balance of payments, unemployment and living standards.

Forecasts are therefore no longer the chief ingredient of energy policy. Nevertheless, insofar as they reflect, or even define, the problems that policy makers are addressing, they can exert an important indirect influence on decision making. It is worth comparing official forecasts made in 1976 with those made in 1978 in order to illustrate important shifts in government thinking.

The National Energy Conference in June 1976 (the forerunner of the Energy Commission) provided the occasion for a number of forecasts to be given a public airing. The forecasts pointed to an 'energy gap' due to strong demand growth combined with rapid depletion of North Sea oil and gas. They indicated that, without major investment in nuclear power and coal, Britain would have to rely on increasingly large and expensive oil imports after about 1990. None of the scenarios examined delayed this prospect for more than a few years. The forecasts were challenged, among the fuel industries, only by the British Gas Corporation (BGC), which thought that North Sea gas supplies would last beyond 2000. To the other fuel industries, the forecasts implied expansion, no difficult political choices between the various fuels, and therefore a secure future. However, the forecasters made no attempt to explain where and how the postulated large additional energy demands would arise in the economy, nor to justify the underlying assumption of uncontrolled depletion of North Sea oil.

For coal and nuclear power, in particular, the accent in official thinking was on expansion. Coal's target capacity for 2000 was set at 170 million tons—the maximum attainable given the need to replace a large proportion of current capacity (due to high costs or impending exhaustion), the long lead times for Selby and Belvoir, and the impracticability of starting numerous big projects simultaneously.

Much the same climate encouraged the UK Atomic Energy Authority (UKAEA) to present, in its evidence to the Royal Commission on Environmental Pollution, a 'reference case' programme which envisaged 104 GW of nuclear plant installed by 2000—including 33 GW of fast reactors[2]. The forerunner of the fast reactor programme was to be a 1300 MW demonstration plant, originally code-named CFR-1 and later CDFR. The reference programme would require at least 6 GW of nuclear plant to be commissioned each year after 1985—that is, installing as much each year as the whole AGR programme which began in 1965 with the ordering of Dungeness B.

Compared with the 1976 forecasts, the changes revealed by the 1978 forecasts are striking. The 1978 Green Paper on energy policy contained much lower demand estimated for 2000—a value of 450–560 mtce (million tons of coal equivalent)[3]. For comparison, the *midpoint* of the official 1976 forecasts was 558 mtce and the top of the range was 760 mtce[4]. The 1978 forecasts marked not only a major downward revision in government estimates but also the overthrow of the 'energy gap' philosophy which was previously held to justify big expansion programmes.

As it happened, the 1978 official forecasts were very close in several respects to those published at almost the same time by Chesshire and Surrey[5]. The two sets of estimates are very close for final energy consumption (in both the 'high' and the 'low' case) and for primary energy consumption in the 'high' case (see *Table 2.1*). But in the 'low' case, the Green Paper estimate of primary energy demand was 115 mtce higher than ours. Since the estimates of final energy consumption are so close, and because non-energy uses (mainly feedstocks) account for only a small part of the difference, it follows that the bulk of the difference stems from different views about the growth of electricity consumption (and hence conversion losses). Evidently, the Department of Energy was assuming much stronger growth in electricity demand in their 'low' scenario than we considered to be plausible.

This illustrates, I think, that the future growth of electricity consumption is a major uncertainty in energy policy. Given their dependence on electricity generation, this uncertainty is of key importance for nuclear power and coal. Since we are here concerned with nuclear power, I shall dwell on that aspect of the uncertainty.

Until it is established that nuclear reactors can work satisfactorily—from a technical as well as an economic viewpoint—on a load-following, intermittent basis, nuclear power will be used predominantly to supply base-load demand. Its growth will be governed not by simultaneous maximum demand on the system, but by the size of the base load, or the continuous demand throughout the year. For the next 10 years the surplus of plant on the CEGB system, combined with the large amount of plant still under construction, means that only a small amount of new base-load plant can be ordered.

24

TABLE 2.1 1978 FORECASTS OF UK ENERGY CONSUMPTION IN 2000

	1977 actual	Forecasts for 2000	
		SPRU*	Green Paper†
	(billion therms)		
Domestic	15.0	14.2–16.2	(13.6)–15.8
Transport	13.1	17.2–19.9	(15.8)–17.0
Other final consumers	7.6	6.0– 8.7	(7.0)– 9.0
Iron and steel	4.9	3.7– 8.4	(7.0)– 9.9
Manufacturing industry	17.9	19.6–28.2	(20.0)–25.4
Final energy	58.5	60.8–81.4	(63.4)–76.8
	(million tons of coal equivalent)		
Coal	122.7	83–162	170
Oil	136.6	124–224	150
Natural gas	62.8	88	50–90‡
Nuclear and hydro	14.3	40–102	95
Primary energy	338.4	335–577	(450)–560

*Chesshire, J. H. and Surrey, A. J., *Estimating UK Energy Demand for the Year 2000: A Sectoral Approach*, Occasional Paper No. 5, Science Policy Research Unit, University of Sussex (1978).
†Department of Energy, *Energy Policy–A Consultative Document*, cmnd 7101, HMSO, London (1978). The figures in brackets apply to a 'low' case which is only partly specified.
‡These are estimated availabilities of indigenous fuels only. They sum to 465 mtce, in the 'high' case, leaving net imports of 45 mtce and a 'policy gap' of 50 mtce.

It might be argued, however, that nuclear power will become relatively cheap and that, as more nuclear plant is installed, electricity will become much more competitive against other fuels. Let us suppose that nuclear power does become relatively cheap (a big assumption) and that it gains widespread public acceptance. At present electricity is very expensive in relation to all other fuels–the average industrial price of electricity in 1978 on a per therm basis was over six times dearer than coal, over five times dearer than gas and nearly four times dearer than oil. True, electricity has certain non-price advantages in many applications. But as long as its relative price is anywhere near so high, it will not substitute for other fuels in competitive markets. That it has retained a market share, despite its very high price, reflects that much of the demand is for non-substitutable and premium uses. Because nuclear power accounts for only a small proportion of electricity supplied (13% in 1977), the price of electricity will not fall appreciably relative to oil and coal until the base load is supplied predominantly by nuclear (still assuming that nuclear power becomes 'cheap').

But that would entail a large nuclear programme to displace fossil-fuel stations from base-load operation, quite possibly the scrapping of many coal-fired stations built in the 1950s and 1960s and, undoubtedly, a big fall in coal consumption. Since two-thirds of coal output currently goes to power stations, it would mean a correspondingly large fall in coal production—making it less likely that sufficient coal output capacity will be available to supply crude heat demand (including substitute natural gas) when North Sea oil and gas supplies diminish and oil imports are very expensive. Further, the CEGB plant surplus and the uncertainty over the relative merits of the AGR (advanced gas-cooled reactor) and the PWR (pressurised water reactor) would increase the costs and the risks of commencing series ordering before the mid-1980s.

One is inescapably drawn to the conclusion that the growth of electricity demand over the next two decades is likely to be no higher than over the past decade (under 3% per annum), especially if natural gas supply climbs to 6000 million cubic feet per day and, as predicted by the BGC, remains at that level until 2000. The scope for electricity growth will be largely determined by the growth of 'premium uses' where electricity has advantages that outweigh its high price. Although this is not the place to develop the argument in detail, the owner-ship of many electric appliances appears to be near the saturation level and the scope for some high-load appliances appears limited. Even if appliance ownership levels increase, design and technical improvements are likely to reduce the average energy consumption of many new appliances[5].

Perhaps the biggest change in official thinking is that the 1978 Green Paper predicts a nuclear component of only 25–40 GW in 2000, which implies the con-struction of only 1–2 GW annually from 1985. The central question for the nuclear industry is no longer whether it can muster the skills and resources necessary to build a very large programme (6 GW per annum), but whether it can maintain itself on a viable basis during protracted low ordering at home and con-tinued difficulty in exporting because of the lack of a commercially proven thermal reactor design.

Following Ince B, Drax B and the two recent AGR orders, further advanced orders for power stations should be out of the question on cost grounds. It is therefore urgent to identify the resources and skills that are specific to nuclear design and engineering and how they can best be organised and retained against the time that series ordering can take place. This problem is connected with the choice of thermal reactor. Clearly, it must take precedence over the fast reactor question, for it would be folly to commercialise fast reactor technology without ensuring that the design and engineering base is there, so that nuclear power stations can be built efficiently when they are needed. If there is a solution, it is likely to require the rationalisation of the heavy power plant industry, including large steam turbine generators, boilers and nuclear design and construction.

Another big change in official thinking between 1976 and 1978 is the recog-nition that the lead times make it impossible to install a large fast reactor programme within 20 years. For example, if CDFR were started in 1981–83 and completed by 1988–93, then, allowing for the necessary operating experience to be gained, the first commercial fast reactors could hardly be ordered before 1990–95 or be in service before 1998–2005. This recognition has brought a cor-responding change in the reason adduced for proceeding with CDFR. While it was seen as the first of a large programme, to be quickly followed with series ordering, it could be held that the cost of CDFR over and above that of alternative power would be recouped from the economic benefits accruing from the subsequent commercial fast reactors. On this line of argument, CDFR would carry no opportunity cost in terms of other R & D that would be foregone. Questions about resource allocation were answered by the claim that the programme as a whole would show a large and positive net present value after applying the test rate of discount, the yardstick for all public sector investment.

Thus, as long as the justification was primarily economic, the important question concerned the estimated cost of fast reactors compared with thermal reactors. This largely hinged upon whether the price of uranium will rise to the extent required to offset the capital cost disadvantage of fast reactors. By 1978, uranium scarcity seemed much less threatening because of major new discoveries (especially in Canada and Australia) and the drop in world-wide nuclear ordering.

However, if approval for CDFR is seen as carrying no commitment for series ordering of commercial fast reactors before the turn of the century, the choice must be seen primarily in the context of R & D (research and development) strategy. It is true that the objective of energy R & D is to open new options, both to increase long-term supply and to reduce the risk of high prices and short-ages of fossil fuels and uranium, and that the ability to build commercial fast reactors would serve this objective. The same is also true of a number of other technologies including, in the nuclear field, high-temperature gas-cooled reactors (HTR) and heavy-water reactors (HWR). HTRs could be developed to supply high-temperature heat for large industrial applications such as chemicals and steel-making; HWRs could be developed to use the thorium fuel cycle, thus reducing dependence on uranium. The advantages of these two alternative advanced reactor systems were heavily stressed until CDFR became the focus of attention; it is important not to lose sight of them as possible options for the future.

The question of R & D opportunity costs looms large if approval for CDFR implies no commitment to early series ordering, for the money to build it must come from somewhere. Something will be foregone. Should the taxpayer or the electricity consumer pay the costs (over and above the costs of alternative power)? If it is not reasonably certain that a highly economic fast reactor programme will soon follow, the question of risk becomes very important with an R & D project that is widely expected to cost £1000 million or more. Given the statutory obligation of the electricity industry to supply electricity at the lowest cost, and given also the pressure to contain public expenditure, neither the CEGB nor the Treasury will be keen to underwrite an open-ended commitment.

On the question of resource allocation, I do not think it holds water to main-tain that a decision to spend such a sum on one R & D project will leave other R & D priorities unaffected. To commit £1000 million or more to one project would inevitably influence the climate of opinion about the alternatives. Directly or indirectly, the funds available for alternative technologies would be reduced. Therefore, before CDFR is approved, we have to be quite sure that an equivalent funding of alternative R & D projects (not only in the energy field) would not be at least as rewarding.

Whether Britain should acquire the capability to build large fast reactors as an insurance against possible long-term uranium scarcity involves a decision which is essentially political. It is political because it requires judgements on the accepta-bility of a whole range of risks (safety, proliferation and economic) and the appropriate rate of social time preference (how much the present generation is prepared to forego for the well-being of future generations). These judgements lie

27

outside the realm of technology and economics: they must be made by politicians.

Almost in parentheses, I may add that the export argument, which occasionally surfaces in the CDFR debate, is wholly spurious. No-one can say whether there will be opportunities 20–30 years hence to export 1300 MW fast reactors. My own guess is that precious few opportunities will arise because of entrenched protectionist policies in the industrialised countries (closed home markets are a strong feature of the power plant industry) and because few developing countries will have power systems to accommodate 1300 MW reactors. On foreign policy grounds, it is surely irresponsible to promote exports of fast reactors, and the plutonium fuel to go with them, until rigorous international safeguards for reprocessing and the large-scale commercial use of plutonium are agreed and implemented.

Also, because of the long plutonium 'doubling time' (25–40 years), fast reactors cannot provide much insurance against uranium scarcity until well into the next century[6]. Yet a further rider is that, because the supply of plutonium is governed by the build-up of thermal reactors, it does not necessarily follow that a 10–20 year postponement of the commercial introduction of fast reactors will reduce their potential contribution to energy supply. The bigger the stock of plutonium when commercial fast reactors are introduced, the more rapid is the feasible build-up of fast reactor capacity.

Assuming that the political judgement is that Britain should acquire the capability to build commercial fast reactors, the relevant question then concerns the best way of proceeding. Building CDFR is only one of the options for acquiring this capability. There are three further sets of questions concerning R & D strategy which have yet to receive an answer.

Firstly, why is it necessary to scale up to 1300 MW? Does not the ability to replicate and perhaps stretch the 250 MW Dounreay prototype fast reactor design give sufficient insurance against the risk of uranium scarcity? The operating performance of very large nuclear units (over 1000 MW) is far less satisfactory than that of medium-sized reactors; large nuclear plants seem especially prone to difficulties with the steam generator and the turbine generator. Practical operating considerations suggest that a size of 500–700 MW would be less risky. Although I have yet to hear of them, if there *are* compelling technical reasons to undertake development work to permit an eventual scaling up to 1300 MW, why is it not possible to concentrate R & D effort on the specific, critical engineering and metallurgical problems rather than building a 1300 MW power station?

Secondly, since France and Germany are both intent upon building demonstration plants which are broadly similar (though not identical) to CDFR, why not wait and license from them when their designs are proven? After all, many countries have acquired the ability to build light-water reactors on the basis of foreign licences, whilst Britain has spent untold sums in developing indigenous thermal reactor designs. Alternatively, why not collaborate with Germany, learning through participation so as to be able to incorporate any necessary design modifications in an eventual British version? The argument that Britain must continue to go it alone because we have little to offer in such collaboration hardly squares with the oft-repeated claim that Britain leads the field in fast reactor technology or with the fact that Britain (unlike Germany) has several years experience in operating a prototype fast reactor. If French or German chauvinism is the obstacle, it might be appropriate to seek a collaborative arrangement with

28

the USA or Japan, rather than concluding that 'going it alone' is the only solution.

Thirdly, given the decisions to build two further AGRs (apparently involving extensive design modifications) and to undertake the design work necessary to enable the CEGB to order a PWR in or after 1982, will it be possible to proceed with CDFR without seriously overstretching the limited design and engineering resources of the nuclear industry? This brings us back to the central problem concerning Britain's nuclear design and construction resources. I fail to see how the industry's problems will be solved by burdening it with three major reactor design programmes simultaneously—especially if it coincides with a further difficult round of restructuring and rationalisation. From an engineering view-point, the risks are daunting.

So far, attention has focused on the question of whether CDFR should be built; but it is worth paying attention to what happens if it is not built. As fast reactor work accounts for two-thirds of the scientists and engineers employed by the UKAEA on reactor development, it is reasonable to conclude that a decision not to build CDFR will immediately call into question the future of the UKAEA. This question may be postponed if CDFR is built, but it cannot be ignored indefinitely. Sooner or later, it must face any single-mission R & D institution. It has already been faced in the USA, where the Atomic Energy Commission was amalgamated into the Energy Research and Development Administration and now into the US Department of Energy. In Canada, the Royal Commission on Electric Power Planning, under Dr Porter, has recommended that the Canadian nuclear R & D establishment should become an energy R & D agency.

It is true that the UKAEA has partially diversified into ancillary work (with considerable success); but I would like to think that its skills and resources, which are unique in Britain, can be applied on a much larger scale to other technologies with the same brilliance and dynamism that were applied to nuclear technologies in the 1950s.

This is not to suggest that the only way forward is for the UKAEA to become an *energy* R & D agency. The decision to change its role must be taken in the context of overall industrial and R & D strategy and the need to make the best use of all our public laboratories. The resources of the UKAEA constitute a major national asset. The assurance of a role which is nationally and personally worth-while will remove much of the uncertainty for the scientists in question. If nothing else, it will make it less likely that Britain builds a 1300 MW fast reactor for the wrong reason.

REFERENCES

1. Cook, P. L. and Surrey, A. J., *Energy Policy: Strategies for Uncertainty*, Martin Robertson Ltd, Oxford (1977)
2. UKAEA, Evidence Submitted to the Royal Commission on Environmental Pollution, Table 3, p. 12
3. Department of Energy, *Energy Policy—A Consultative Document*, cmnd 7101, HMSO, London (1978)
4. Department of Energy, *Energy R & D in the United Kingdom: A Discussion Document* (1976)
5. Chesshire, J. H. and Surrey, A. J., *Estimating UK Energy Demand in the Year*

29

2000, Occasional Paper No. 5, Science Policy Research Unit, University of Sussex (1978)
6. Grainger, L., 'The nuclear issue as seen by a competitor', *Energy Policy*, Vol. 4, December 1976

3

The Electricity Sector and Energy Policy

Peter Odell

3.1 INTRODUCTION

In the context of an analysis of energy requirements and the fast breeder programme, one of the main considerations must be the prospects for the energy
economy's electricity sector. It is, of course, this sector that the nuclear
programme is essentially designed to serve, so that expectations for electricity
will be the main element which shapes the size and speed of nuclear development,
in general, and within the framework of which the fast breeder programme, in
particular, must be determined. However, the growth of the electricity sector is
not simply a function of the rate of growth of the total demand for energy in
the economy. It is also related to institutional decisions on the issue of what may
be called 'the electrification of society'—a concept which is sometimes explicitly
stated[1] but one which is much more often implicitly assumed as part of the very
widely held view that the demand for electricity in any economy is bound to grow
at a rate in excess of energy demand in general. Indeed, such a development often
seems to be held to be a 'law' of the energy economy and one, moreover, which is
widely believed to be eminently desirable. As a consequence, the whole question
of the relationship between the demand for electricity and the alternatives for
energising society has become a largely unexamined part of energy policy making[2].
It is to this issue that this paper is directed, as the costs, safety, security and other
problems of the fast breeder programme mean that we cannot afford to allow the
issue of the electrification of society to remain an unexamined side issue in energy
policy decision taking. On the contrary, I want to suggest that it ought to be one
of the main variables for analysis prior to the fast breeder programme decision.

3.2 THE CONCEPT OF THE ELECTRIFICATION OF SOCIETY

(a) The term 'the electrification of society' is a short-hand way of expressing a
belief in the inevitability and/or the desirability of developing an energy economy
in a way in which electricity comes to meet an increasing share of energy end use,

31

including the provision of heat in homes, offices and factories and of the energy required in transportation. As society becomes richer, it is argued, so its members demand the cleanliness and convenience of electricity in preference to other ways of meeting their energy needs[1]. Thus, increasing per capita income carries with it the inevitability of relatively higher growth rates for electricity demand than for non-electric energy. This, indeed, is a phenomenon which has been noted in all parts of the western world over the last 20 years. The general validity of the hypothesis is, moreover, indicated by the relatively higher use of electricity in the countries with the highest per capita incomes. In other words, it seems to be a special law within the general law concerning economic growth/energy use relationships. Paradoxically, however, whilst the latter has taken something of a beating in the context of the much higher real energy prices of recent years and because of fears of a 'scarcity' of energy resources, belief in the special 'electricity use law' seems to have strengthened. This is because the now very generally accepted energy-gap-filling role of nuclear power implies a greater emphasis, even more quickly than hitherto anticipated, on the use of electricity which, with present technology and with constraints on the location of nuclear power stations, is virtually the only means whereby nuclear power can be delivered to consumers in a usable form.

(b) In most western industrialised societies, including the UK, the pre-existing base from which to take the necessary steps towards the further electrification of society is an electricity supply system which emerged as a response to the elect-ricity needs of low-electricity-use energy economies. These economies have mainly had their relatively limited electricity demands met by low-conversion-efficiency thermal power stations using coal or oil as the input fuel. These, for reasons of economies of scale, in respect of both the plant itself and of the associated trans-port facilities, and also for reasons of an environmental nature (neither coal nor oil-fired power stations of upwards of 500 MW are 'good neighbours' in urban communities), have generally been located somewhat remotely from energy demand centres, so that the possibilities of utilising the waste heat from the power stations have been almost non-existent. Moreover, also as a consequence of power station location patterns, high-voltage and often long-distance transmission lines (also considered to have considerable adverse environmental impact) have been required to link production with demand centres. They thus also form an integral part even of the relatively low-electricity-use energy economy from which the more intensive electricity economy has to be developed with a further requirement for an intensified system of high-voltage transmission lines.

3.3 THE NECESSARY CONDITIONS FOR PROCEEDING TOWARDS THE MORE INTENSIVE ELECTRICITY ECONOMY

(a) The system described in 3.2(b) above needs a continuing availability of suf-ficient low-cost fossil fuels to enable it to continue to deliver decreasing real-cost electricity so that the demand for electricity can be kept moving steeply up. Even if the most favourable condition of declining real-cost availability of electricity cannot be met, then it is at least necessary for the price of electricity to decline relative to the price of fossil fuels delivered to final users so that consumers, particularly in the residential and commercial sectors, will prefer to use electricity for space-heating uses. Then, as a result of the continuing rapid increases in electricity demand, accompanied by the achievement of improving load factors on the system arising from the ability to sell cheap off-peak electricity for residential and commercial space-heating loads, commercial conditions are created for the electricity supply authority which are appropriate for the initial intro-duction of nuclear power stations into the system.

(b) From this moment, the second necessary condition for proceeding towards the more intensive electricity economy can be met. This is the development of large-scale nuclear power production into the system in order to achieve an availability and use of generating capacity which is large enough to create 'popular' acceptance of the idea that the steady and continuing expansion of the electrification of the energy system is self-evidently true. Thereafter, a situation of self-sustaining growth will be achieved through economies of scale in both production and in distribution, with consumers securing access to electricity at decreasing unit cost and so being encouraged to convert gradually—and inevitably—to all-electric energy systems covering the domestic, commercial and industrial sectors of the economy.

3.4 IMPLICATIONS OF RECENT BASIC CHANGES IN THE ENERGY MARKET FOR THE 'ELECTRIFICATION OF SOCIETY' PROCESS

(a) In 1971, the long period of almost 20 years of falling real-cost energy came to an end because of the fundamental change in the international oil market arising from the December 1970 decision of the Organisation of Petroleum-Exporting Countries (OPEC) to assume control over the supply and price of internationally traded oil and the cooperation of the oil companies in this changed situation[3]. Since then, and more especially since 1973, the real price of oil and of other fossil fuels has increased sharply (in 1979, the real price of oil is over six times higher than it was in 1970 and twice as high as in 1950). Insofar as most of the western industrial world's electricity systems in the late 1970s still remain highly exposed to the impact of such changes in the cost of fossil fuels (given their general ability to utilise little more than 30% of the thermal value of the fossil fuels they burn), there has been a consequential very sharp rise in the price of electricity. This, coupled with the expectations and perception of consumers that price increases in the electricity sector will now continue, has not only led to a significantly diminished rate of growth in the demand for electricity, but also to much resist-ance on the part of consumers to commit themselves to decisions which would serve to increase their consumption of electricity in the future. The hitherto expected continuing high rate of increase in the sales of kilowatt-hours has failed to materialise and has thus produced cash-flow problems in relation to the hither-to planned self-financing expansion of electricity systems based on highly capital-intensive nuclear power capacity and associated transmission networks. This unfavourable position for expansion into nuclear power has been exacerbated by the creation of unexpected surplus capacity in the electricity systems as a result of the post-1973 completion of long-planned generating capacity which was originally scheduled against the expectation of continued high growth rates in the demand for electricity. This has a two-fold effect: first, as the high fixed costs in the electricity supply and distribution system have to be charged against lower-than-expected unit sales, electricity price increases to consumers are enhanced even beyond those caused by fuel price rises. Secondly, given the existence of surplus generating capacity in the system, the ability to argue successfully for the 'need' for new nuclear capacity is diminished.

Even ignoring the last point (as appears to be the tendency of many electricity system planners on the grounds that the recent slowdown in the rate of increase in consumption is merely a temporary phenomenon), there is, nevertheless, still a 'Catch 22' situation over the future outlook for the electricity sector, namely, that if electricity prices are kept low so as to keep consumption rising, then insufficient funds can be generated—or borrowed—to finance the expansion, but if electricity prices are increased to improve the investment outlook, then growth in consumption is so restrained as to make massive nuclear-based expansion of the system unnecessary. In other words, the second necessary condition (3.3(b) above) for proceeding towards the more intensive electric economy cannot be met.
(b) Apart from the economic problems now associated with the expansion of the electricity system based on nuclear power, there is also the recently much accentuated opposition to the nuclear power option for meeting future energy needs. This is aimed not only, or even mainly, at the nuclear power stations themselves, but also at various elements in the fuel-cycle system which is a necessary attribute of nuclear electricity production. This non-acceptability of near-future nuclear power amongst some sections of the population also serves to undermine the base from which the much more intensive electric economy of the late twentieth century was to be developed.

3.5   THE PRESENT-DAY DESIRABILITY AND INEVITABILITY OF THE ELECTRIFICATION OF SOCIETY

However, in spite of the developments outlined in section 3.4 above, the 'desirability' and 'inevitability' of continuing progress towards the electrification of society still appear to be presented as virtually non-arguable elements in the energy debate. The following points seem to be worthy of attention in these respects.

3.5.1   The Desirability of Electrification

(i) This, as far as energy users are concerned, seems unlikely to be independent of price. In current energy market conditions, any proportional increase in the use of electricity by a consumer will normally raise his energy costs overall. Thus, even if we assume that electric space and water heating are alternatives in the commercial and residential markets, would users accept the need to pay a premium price for it? And if so, then how much of a premium? Not very much of one it seems, if the recent necessary reduction for marketing reasons in the price of off-peak electricity for domestic heating purposes may be taken as demonstrating good evidence of consumer resistance to the use of electricity when cheaper non-electric alternatives are available. And even relatively high resource cost develop-

34

ments of off-shore oil and natural gas still provide cheaper non-electric alternatives for most energy consumers[4].

(ii) As far as the 'public at large' is concerned, the paraphernalia of electrification (the power stations themselves, the transmission lines, the switching/stepping-down stations) are visually intrusive and are certainly not seen as desirable unless it can be shown that the alternatives for energising homes, offices and factories etc. are worse. And this is clearly not so in a situation in which natural gas can be made available to these consumers without any environmental problems worthy of note.

(iii) As far as 'government' is concerned, there seems to be no inherent reason why electrification as such should be desirable: except in the very special case of all alternative energies being so scarce that a country's future ability to survive depends on nuclear-based electrification. This is obviously not the case in respect of the UK with its very considerable known and/or probable reserves of fossil fuels. In all other cases, the desirability of the increasingly intensive use of electricity must be a function of the comparative costs and benefits of electrification in relation to all other alternatives.

3.5.2 The Inevitability of Electrification

(i) In the period before there was much concern for the future availability of fossil-fuel resources, the argument for the inevitability of electrification was basically of a simple technological–developmental sort: 'If it is new, then it must be cheaper/ better than alternatives which are older and/or out of date'. Nuclear power had a special place in the context of this sort of argument—for two reasons. First, because the funds for its development seemed to be essentially unlimited and, secondly, because its development could be placed in a 'national pride' slot.

(ii) More recently, with the evolution of the inherent scarcity-of-energy syndrome, it has been possible to add an apparently unassailable argument to justify nuclear-based electrification, namely, that only nuclear power can meet the world's needs for energy. Note, however, in this argument that the potential role of demonstrably non-scarce coal is glossed over and that oil (and gas) are simply assumed to be in short supply in a situation in which there is much evidence that there will be a ready availability of these commodities for the whole of the medium-term future (into the second quarter of the twenty-first century) at the levels of real prices already achieved. The assumption of scarcity is valid only for specific political and institutional conditions, and there seems to be no reason why we should assume that the present unfavourable conditions in these respects will always be with us[5].

(iii) Note, moreover, that the belief in a scarcity of fossil fuels arises out of the particular methodology used to calculate the future demand for energy. In the absence of nuclear power, it is assumed in the approach used, fossil fuels will be needed in sufficient quantities to make as much electricity as will be produced in a nuclear-based electricity-intensive economy. Moreover, the calculation of nuclear electricity's contribution to the energy economy is made on a fossil-fuel equivalence base (that is, through calculating the amount of coal (or oil) that would be needed to make that amount of electricity, assuming an approximately 35% fuel conversion efficiency in conventional power stations), rather than on the heat-energy equivalent basis of the electricity produced*. The effect of this approach is to exaggerate by almost three times the apparent contribution of nuclear power to the total amount of energy consumed. It also has the effect of apparently increasing the size of the expected demand for energy overall, so that without the development of nuclear power there is an exaggerated apparent demand for 'scarce' fossil fuels.

(iv) This methodology used for stating the expected evolution of energy demand implicitly assumes that electricity is not a variable component in the energy mix. The assumption, in other words, is that all future demands for electricity are non-substitutable. By implication, this means that it is thought that none of electricity's expected increased contribution to the future energy economy can be provided directly by fossil fuels used in inherently more efficient ways. But, as much of the increased availability of electricity will be used for providing heat energy and/or motive power, this is clearly invalid and so raises doubts over the validity of the methodology used. Indeed, the approach to the method of evaluating the energy contribution of nuclear electricity and of indicating the degree to which it is substitutable is an important one. It is certainly more than a narrowly technical question over the choice of conversion factors. It is perhaps not without significance that the official method chosen for calculating future energy demand helps in itself to suggest the inevitability of the electrification of society.

3.6 ALTERNATIVES TO THE ELECTRIFICATION OF SOCIETY

(a) There are several important questions in respect of the nature of the future demand for energy which need to be looked at, even if one first makes the assumption that no growth or near-zero growth in energy use in economies such as that of the UK are not reasonable options for the medium-term future. This assumption itself has, of course, been challenged in recent studies[6].

(i) Is it 'desirable' and/or 'possible' to minimise all 'non-necessary' uses of electricity? Could the pricing mechanism be used appropriately to achieve such an end? Or are controls over consumer choice necessary to achieve that end and, if so, are they likely to be acceptable and effective[7]?

(ii) To what degree and at what cost is it possible to develop a geographically dispersed pattern of electricity supply in which the electricity is a joint product of, or a by-product from, heat production? This implies more than the consideration of district heating systems (though these need to be reconsidered with broader terms of reference than those reported in Energy Paper No. 20 of the Department of Energy[8]). Quicker and, perhaps, even more significant results in respect of dispersed electricity production may well be possible with the widespread installation of 'total energy' systems in individual industrial and commercial premises or for groups of such users' premises and for public-sector facilities such as hospitals, administrative centres and swimming baths, etc. How can such entities and areas for combined heat and power developments be identified? Is sufficient statistical or cartographic information available to enable identification to take place? To what degree would present law and existing institutions have to be changed to encourage or to make such developments possible? Ought the planning processes

*See Appendix 2.

to include a requirement for alternative energy evaluations of all proposed developments?

(iii) Is interconnection with the national electricity grid feasible for such total energy systems in large numbers? And at what cost? Can the example of the Midland Electricity Board's initiative in developing an industrial combined heat and power (CHP) facility at Hereford be repeated ten, a hundred or a thousand times? If so, with what consequences as far as the need for new central power stations is concerned? If interconnection of such local systems with the grid is not possible and/or acceptable, then what is the significance of a large number of industrial/commercial CHP systems for the future viability of the public electricity system?

(iv) What is the current technico-economic status and what are the socio-political implications of individual household natural-gas-based fuel cells with heat recovery potential (and possibly linked to heat pumps and/or solar panels) for those consumers who are not geographically well sited for incorporation into a district heating system? Would such developments provide an additional, economically attractive load for the natural-gas supply industry, given its expectation of a rising availability of gas for the rest of the century[9]? Would the capital cost involved be greater or less than in providing nuclear-based electricity production to meet the same demand for energy? And what are the contrasting labour input requirements of the two alternatives?

(b) For the short term, the encouragement of dispersed electricity production with waste heat recovery may not necessarily be appropriate, because existing or firmly planned centralised electricity generating capacity can meet anticipated demand for a number of years ahead (to 1987?). Nevertheless, the introduction of dispersed CHP facilities on a much larger scale would seem likely to lead to energy conservation on a large scale and it would also enable the electricity authorities to retire their most inefficient or most inappropriately located producing facilities. Can the costs and benefits of the alternatives be quantified and can optimal degrees and patterns of CHP be established?

(c) For the medium to longer term, what constraints are there on continuing CHP expansion? What, for example, are the implications for natural-gas use and for natural-gas depletion policy of, say, an annual installation of 1.5 GWe in gas-fired CHP systems (to give about the same addition to useful energy as the 3 GWe of nuclear electricity planned for each year after 1985). If there are oil and gas resource base constraints in such a development, then could these be overcome by using coal instead with the application of fluidised-bed combustion technology?

(d) If the long-term energy future (say, post-2020) will have to be based on fission/fusion technology, involving highly centralised producing systems, is it worthwhile developing an interim dispersed pattern of combined electricity and heat production? In other words, are the net economic benefits in the medium term from such a development sufficient to justify the longer-term dislocations which could arise if and when a return has to be made to centralisation? On the other hand, however, one should note that dispersed electricity and heat production facilities, implying a minimum electricity path and based essentially on oil and gas as input fuels, may serve to stimulate the hydrocarbons supply potential— in terms of both conventional supplies and supplies as synthetic natural gas (SNG) and/or oil from coal. If so, then the required date for re-establishing a centralised system may be put back indefinitely. Or, given a medium-term dispersal of electricity production, may this not stimulate longer-term non-centralised developments based on alternative technologies which are sufficient to meet a minimum electrification path?

3.7 CONCLUSIONS

(a) The more intensive electrification of society, as hitherto generally expected and as still implicitly planned in official views of the energy future, is, however, now only one possible option out of several which seem able to make increasing quantities of useful energy available to energy consumers of many different types. All the options are open in the context of an assumption of modestly increasing energy use (at, say 3% per annum). The different options involve wide variations in the contribution of electricity to the total amount of energy used in final consumption and hence imply contrasting levels of demand for primary energy[4].
(b) Moreover, the idea of the inevitability of increasing energy use is itself being challenged, specifically in respect of industrialised economies such as those of the United Kingdom[6]. Given an average rate of increase in energy use of less than 2% per annum in the western industrialised world, then the energy 'problem' arising from doubts over the availability of oil and gas is put well back into the twenty-first century. This is true largely irrespective of what Third World countries decide to do (or can do) in the meantime as far as their energy use is concerned. It is, however, important to remember that this latter group of countries' behaviour is very dependent on the demonstration effects of the processes which are introduced first in the industrial countries. The electrification of society in the latter group of countries will ultimately tend to lead to the same trend in the evolution of the energy economies of the technologically dependent Third World to which there will inevitably be a diffusion of decisions similar to those taken in the West. This means that the decision that still remains to be taken by the UK and other industrialised countries—that is, the choice between a nuclear-based electricity-intensive economy, on the one hand, or the largely non-nuclear, minimum-electrical-energy economy, on the other—is of great significance as far as international aspects of contemporary energy issues are concerned. In this context, appropriate research into the fundamental questions of energy/electricity supply and demand is an essential part of energy policy making. It is surely self-evident that research into these issues should be undertaken before the decision on the fast breeder programme is made.

REFERENCES

1. Brookes, L. G., 'Towards the All Electric Economy', *Energy: from Surplus to Scarcity*, Applied Science Publications, London (1974)
2. Department of Energy, *Energy Policy: a Consultative Document,* HMSO, London (1978)
3. Odell, P. R., 'The World of Oil Power since 1973', *Oil and World Power*, 5th edn, Penguin Books, Harmondsworth (1979)

4. Odell, P. R., 'Europe and the Cost of Energy: Nuclear Power or Oil and Gas?', *Energy Policy*, Vol. 4, No. 2, June 1976
5. Odell, P. R. and Vallenilla, L., *The Pressures of Oil*, Harper and Row, London (1976)
6. Leach, G., Lewis, C., van Buren, A. and Romig, F., *Low Energy Scenario for the UK, 1975–2025*, International Institute for Environment and Development, London (1978)
7. Chapman, P., *et al., A Critique of the Electricity Industry,* Open University Research Report, ERG 013 (1976)
8. Department of Energy, *District Heating Combined with Electricity Generation in the United Kingdom*, Energy Paper No. 20, HMSO, London (1977)
9. British Gas Corporation, 'UK Energy Prospects and the Plans of British Gas', *National Energy Conference*, June 1976

4

A Low-Energy Growth Alternative

Gerald Leach

The results of a detailed, sectoral low-energy strategy for the UK have recently been published[1]. I shall use some information from this study to take further one of the central issues already raised here, notably in the contributions by John Surrey and Leslie Grainger. This issue is the rapid widening of the energy and electricity demand futures that now seem plausible. The relevance to the need for a nuclear or a fast breeder programme will become obvious.

Let me start by illustrating a point made by John Surrey that the official forecasts of energy demand have been falling very rapidly. The Department of Energy forecasts for the year 1990 have been falling steadily since 1975 at a rate which, if projected, suggests that in 1980 they will be at zero energy growth. In the last two years, the Department's forecasts for 2000 have dropped by 20% in the low economic growth case and 26% in the high case.

I shall return to these forecasts later. Before doing so, I shall look in some detail at the electricity demand forecasts and the extent to which they do—or do not—incorporate conservation and saturations of demand. Electricity is, after all, what a breeder programme will provide, unless anyone is seriously thinking of siting reactors close enough to urban areas to utilise the waste heat. The 1978 Green Paper assumes electricity growth rates of approximately 3% per annum, following the 7% and then the 5% of the last 20 years. There is little validation for that figure, and nothing about the policy assumptions that lie behind it. It seems to be derived from grossly aggregated econometric forecasts using data from the 1950s and 1960s; in other words, pre-oil-crisis data.

Let us, instead, look at recent trends by sector of electricity consumption, as shown in *Figure 4.1*. In the largest sector—housing, with 40% of total consumption—demand has fallen despite three years of colder-than-average winters and a rising number of households. In other sectors, growth has been linear or zero. Now, of course, there are sectors where electricity can and probably will grow, but it is the crude use of extrapolative econometric forecasting that produces, from trends like this, a future of exponential growth at whatever rate per annum: a practice that has been largely responsible for our present power station over-capacity, which began to build up well before the 1973–74 oil crisis.

41

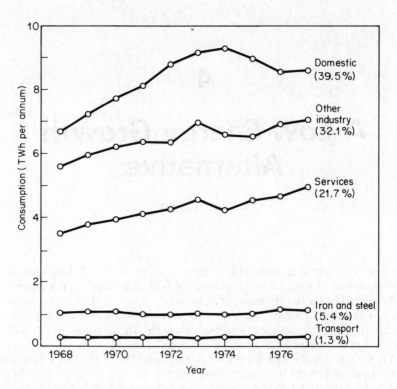

Figure 4.1 UK electricity use by sectors: 1968–77.

To see where electricity might grow in future, consider the crude breakdown of 1976 energy use in the UK by end use, as shown in *Table 4.1*. Only this year has it become possible to construct such a table with any accuracy. These figures are based on a 140-sector breakdown of energy use in 1976 that we have used in our study, including an analysis of the eight major industrial sectors by four temperature bands for direct process heat, four for indirect process heat (steam), space and water heating, electrochemical processes, stationary machines, lighting and off-road vehicles. (The analysis was made by the Energy Technology Support Unit in collaboration with The National Industrial Fuel Efficiency Service using Energy Audit and Energy Thrift Scheme data, so far unpublished.) From this work one can see, for the first time, where energy is actually used and where it can most easily be saved, so that one can set up targets and campaigns for conservation.

From this breakdown, one can obviously exclude the growth of electricity in most of the non-energy market—feedstocks, bitumen and industrial spirits— because they require carbon. To use primary electricity to produce carbon-based feedstocks calls for mining vast quantities of limestone, which will almost certainly be environmentally unacceptable.

Electrification of transport depends on battery development and applies plausibly only to cars, vans, buses and to some extent rail, which is a very small

42

TABLE 4.1 UK ENERGY CONSUMPTION (%) BY END USE IN 1976

End use	Energy source					
	Solid	Liquid	Gas	Electricity	Heat	Total
Heat under 80°C	8.2	10.5	12.5	3.1	1.0	34.8
Heat over 80°C	7.4	7.5	6.5	1.1	2.5	25.0
Essential electricity				8.1		8.1
Transport	0.1	20.8		0.2		21.1
Non-energy uses		9.5	1.5			11.0
Total	15.7	48.3	20.0	12.5	3.5	100

energy user. The complete electrification of this traffic at today's levels by advanced technologies such as the sodium–sulphur battery, with the devices that would be used in electric cars such as regenerative braking, would reduce the total energy demand of transport to about one-fifth of the figure shown in *Table 4.1*, or about 4%. However, such a total electrification will only be possible when there is the capability for rapid refuelling *en route* during journeys; for example, using the zinc–air battery in which the electrolyte can be rapidly exchanged. Transport electrification thus depends on many technical problems that still need to be solved. Furthermore, the electricity that would be used by transport would mostly be stored and would, therefore, smooth daily loads. Although this favours the economics of all kinds of base-load electricity, it also favours intermittent sources such as wind and wave.

The 'essential electric' sector shown in *Table 4.1* consists of lighting, machinery, electrochemical processes, and so on. While it is likely to grow, for a really major expansion, electricity obviously has to penetrate the dominating heating market. At present, the cost and price differentials between electricity and competitive fuels make this most unlikely. At the same time, a range of technologies such as electric heat pumps will soon be widely available for greatly reducing any potential electricity growth in this sector. Similar technical advances, all of them spurred by the events of 1973–74, make other sources of heat—particularly coal and, to some extent, gas—even more attractive than they already are as competitors to electricity in heating. It is not for nothing that the most vociferous advocates of really high insulation levels in housing are the Electricity Council. It is also worth noting how this breakdown reveals the silliness of dismissing solar energy as a poor source of electricity (rather than as a direct provider of heat)—as in Mr Jones' contribution.

Now let me take this breakdown further and ask in greater detail how electricity is actually used in the UK. *Table 4.2* gives a rather aggregated breakdown. Starting from the top, space heating is mostly in housing but also in offices. Even the briefest scan of the conservation literature shows that this electricity use can and probably will for simple economic reasons be greatly reduced even if the stocks of housing and offices increase. On our projections, the insulation of lofts

TABLE 4.2 END-USE STRUCTURE OF UK ELECTRICITY CONSUMPTION (%) IN 1976

End use			Total
Space heating:	domestic	9.9	
	commercial, etc.	3.2	14.0
	industry	0.9	
Water heating:	domestic	6.8	
	commercial, etc.	2.4	9.2
Industrial process heat		8.9	8.9
Lighting:	industry	3.7	
	commercial, etc.	8.8	14.8
	domestic	2.3	
Domestic:	appliances	13.0	
	cooking	5.9	18.9
Machinery:	industry	23.3	
	commercial, etc.	4.1	27.4
Electrochemical processes		4.8	4.8
Transport		2.0	2.0
Total			100.0

and of cavity walls (where they exist) reduces the useful energy for space heating in the existing building stock by 40% over the period 1976–2010. This does not seem to be an immoderate assumption. With new houses and offices, many have been built recently with energy requirements from one-half to one-third of the national average with investment cost penalties that vary from a few percent plus to several percent below the norm. These negative cost penalties occur because many of the conservation options are free—it is just a question of good orientation, siting, and so on—while heating, ventilation and lighting systems can be smaller. The architectural profession is now pressing on receptive ears in Whitehall to tighten strongly the building regulations in housing, with a 50% reduction in heating needs considered a reasonable long-term target even for traditional construction methods.

With these measures and the gradual replacement of the existing housing and office stocks by new and more efficient units, even with substantial increases in living temperatures, we project a substantial fall in heating demand. With water heating, one has very similar opportunities, while quite small increases in the per capita use of hot water will bring us up to American levels of demand. Again, energy use for this purpose could well fall.

Savings in industrial process heat depend largely on pricing and policy, but the expansion of electricity in this sector is not as likely as it is for gas in the medium term and for coal in the longer term, simply because of relative pricing. Lighting is a very significant sector and again, if one looks at the technical literature, very large potential savings appear to be possible. I will just mention two: there is an eight-fold difference in energy use between tungsten and modern fluorescent and mercury halide lights. In the USA, high-frequency fluorescent lighting with a 30% reduction compared to conventional fluorescent lights is now coming onto the market. Many offices in Britain have, by extremely simple measures, saved 50–60% on their lighting load in the last few years with economic payback times of 12–18 months.

The next electricity-using sector—home appliances—is particularly important in terms of peak loading. With cookers and the major electricity-using appliances in the home, and to some extent in offices, a 50% or greater reduction in electricity use is technically possible simply by better insulation, better door seals, slightly better heat pumps, and so on. There are also likely to be strong saturation effects in ownership.

I shall illustrate this point by a study, as yet unpublished, by Art Rosenfeld at the Lawrence Berkeley Laboratory. They examined all refrigerators on the market in California and plotted the cost against the energy use. Whereas the most expensive ones are the most energy efficient, there is no marked correlation. They then costed a range of energy-reducing design improvements and found that to reduce average consumption from about 140 to just over 50 kWh per month would cost an extra 40 US dollars, or very nearly 10% of the average cost. If similar savings and costs for deep freezers are included, and if all Californian equipment was changed to these more efficient models, the investment cost would be $750 million but the saving in power station capacity would be 1700 MW with an investment cost of about $3 billion.

Office and industrial machinery, the next item on the list, will almost certainly grow as an electricity market. But it is also likely to become less energy intensive per unit of monetary output through increased automation and the electronic revolution in general. Electrochemical processing—not a very big item—is almost entirely aluminium production where savings of the order of 30–40% per unit of output are already being taken up as we convert from the traditional Hall–Heroult method of production to the Alcoa process. Production is not likely to increase very much since the value added in aluminium will be retained more and more by primary producers of bauxite.

The technical literature on energy conservation is full of instances of this kind, and indeed, the Government and other official institutions are well aware of this. There is also, I believe, a strong and growing commitment to the virtues of energy conservation in this country—and in other countries. But I would also suggest that only a sectoral forecast which examines each different kind of energy end use in some detail can reveal what the future could bring through more efficient energy use and fuel switching, whether this is brought on by higher prices or conservation policies or both. Only such forecasts can reveal the many strong saturation effects in demand, regardless of the more efficient use of energy, that are bound to occur in future, and that seem to be ignored or missed by the forecasting methods on which our present official Department of Energy forecasts are based.

The Department forecasts as used in the Energy Policy Green Paper, use for three major sectors—that is, the domestic sector, offices and other industry, which together account for 67% of delivered energy and 92% of electricity—a simple, linear energy–economic regression equation. Now despite the six-figure accuracy of the equations one should note three things. First, all energy use for different purposes is lumped together as one form of (useful) energy. In housing, for example, there is no differentiation between space heating, cooking, lighting, appliances or water heating. Secondly, the equations are based on the average household income in the case of the domestic sector, or output in the case of the other sectors, using regressions from data during the 1960–73 period in the case of other industry, and 1954–73 in the case of the other consumers. These equations cannot, in principle, tell you at what income or output level houses or offices become so hot that they are uninhabitable. Neither can they say anything about the effects of future policy. Thirdly, they take no account of important physical saturation effects in heating buildings, which accounts for 45% of national energy use. Any architect is familiar with these effects. They are due to the fortuitous heat gains from cooking, lights, appliances, people, and so on.

It is useful to find out if someone understands this crucial point by asking them a simple question. If a house loses heat at the rate of 100 units and insulation reduces that loss to 60 units, by how much does fuel consumption drop? The answer is not 40%, but in average UK conditions something like 55–60%. This is because the fortuitous heat gains initially contribute about 30% of the heating requirement and remain the same after the insulation is installed. A second important saturation effect that is ignored in the Department forecasts is the possible and likely change in the mix of industrial and services production in the future. In the forecast, 'other industry' is treated as one homogeneous unit and no attempt is made to tease out the different industries within that whole. All industries are assumed to have the same growth rate. Since the energy intensity of industry ranges from about 340 MJ/£ of output in iron and steel down to 30–40 MJ in engineering, food and drink and construction and down to about 15 MJ for the service industries, this point is not unimportant. It seems extremely likely that if one assumes substantial gross domestic product (GDP) growth in future, that can only occur if there are very substantial shifts in the nature of industrial production, probably from the primary and energy-intensive end of the spectrum to the secondary and tertiary sorts of output where energy intensity is quite low. With that shift will come, automatically, a substantial lowering of the average energy intensity of the industrial sector which accounts for about 40% of energy use.

A final point is that the Department of Energy do allow for conservation both to reduce the need for useful energy and to improve appliance and process efficiencies. However, the assumptions are mild, they are not validated in any way, nor are they argued, nor are there any explicit policy inputs to explain them. Could it be—and this is not an entirely rhetorical question—that these mild assumptions are due to pressures from above, particularly from some of the energy supply industries?

When these factors that I have been discussing are corrected—particularly the saturations effects that one can begin to see—it looks increasingly plausible that the UK could have very substantial material growth without any increase in energy or electricity demand. Other recent forecasts are coming to the same conclusion. The most notable of these is the large CONAES study (the Committee on Nuclear and Alternative Energy Systems) in the USA whose preliminary forecast is that with a two- to three-fold increase in energy prices, by 2010 the per capita primary energy consumption would be the same as today's or about 20% below current

46

levels. This is very similar to the figure that we are coming out with for our forecast to 2025 for the UK. We have zero energy growth in our high case, where GDP trebles in the long run, and a 25% decline in primary energy in the low case where GDP 'merely' doubles. In both cases, electricity demand is more or less constant from here on. But what has really surprised us in this study is the moderation in the assumptions that one has to make to achieve such a low-energy growth future. Many experts consulted urged us to assume larger savings, greater increases in technical efficiency, more rapid introductions of these to greater shares of the market than we have in fact assumed. To give examples, our assumptions about the energy efficiency of vehicles, of the introduction of CHP, of renewable energy sources are all considerably more conservative than those in the Energy Policy Green Paper or other Department of Energy Papers. This was quite deliberate on our part. We wanted to show what could be done with proved and tried technologies, not by things which have not yet been firmly established in terms of cost or technological feasibility. The energy savings in industry are generally lower than those found to be reasonable by the Energy Audit and Energy Thrift Schemes or in the recent long-range assessment of energy savings made by Shell. Why the difference then? I would suggest that the real reason is that most traditional forecasts make an extrapolation to the year 2000 based on historic rates and then knock that down by an energy saving. What we have done is to start from the bottom up in a physical model, working from useful energy, and build up from there, sector by sector, to primary energy.

Our main findings are stated quite simply. Coal production stays at today's level in the low case to 2025 and rises to 150 million tons by 2025 in the high case. It is a relaxed coal policy. Nuclear expands to 1990 because of the plant in the construction pipeline and the two extra advanced gas-cooled reactors (AGRs) that are on the way. If it expands further thereafter, the coal burn would have to fall *pro rata*. As a base-line case to keep down the number of variants, we have assumed a nuclear ordering rate of 3 GW per decade from 1980 onwards, with the result that nuclear capacity gradually declines as old plant is retired. The choice is, in fact, entirely between rising nuclear and falling coal or constant coal and falling nuclear. We have only a small component of wave with some wind power after the year 2000. With North Sea oil on central estimates of reserves, there is an excess over consumption until 2005 or 2010 in the high and low cases, respectively. With the upper end of the reserve estimates that are currently being used, a small oil gap develops around 2010 in the high case and after 2020 in the low. With North Sea gas, all demands are met on central estimates of reserves until 2025 in the low case. In the high case, a small gap appears around 2015–2020.

From our work, we believe that to achieve this energy future of low risk, all that is necessary is to apply with a commitment a little more vigorous than is being shown today, by government, industry and other agencies, some of the technical advances in energy use which have been made and are still being made in response to the 1973–74 crisis. By saving on supply, one releases enormous investment funds to spend on conservation. In our scenario, the electrical generating capacity that need not be built, in comparison to the Green Paper forecast, is of the order of 50–60 GW by 2000. Assuming roughly £500/kW, that is £25–30 billion saved in investment, or over £1000 million per annum. We would be extremely surprised if that did not pay for all the conservation assumptions that we have made. We have also assumed two major policy thrusts, not difficult ones, which are already beginning to get under way. One is the setting of energy standards or targets in a small range of key goods where there is a large technical potential for improvement. One of these is new buildings: with the building regulations, we already have the mechanism for setting targets. The second is cars and light vans, where the industry seems to agree that there is a large cost-effective potential for improvement, but no single industry wants to go it alone. The third sector consists of the electrical goods, lighting and cookers where, as I have shown, the improvements with quite small cost penalties are dramatic. Again, industry must be helped to advance in line by having government regulation. The second policy thrust is the promotion of better information on what 'best practice' technologies are available for consumers of all kinds. Government departments and professional institutions are beginning to move very rapidly on this front.

Now clearly such a future for the UK does not require, for energy demand or energy-gap-filling reasons, a breeder programme or even much of an expansion in the nuclear programme. The argument then comes back to other issues. Cost is one of these: the relative costs of coal or nuclear electricity, for example, or nuclear thermal and fast breeders. Having seen contributions to this volume, I am not going to get into that. Beyond questions of cost, are much larger unquantifiable issues. The carbon dioxide problem is one of them. Proliferation is another. I would add the whole basket of issues that are to do with scale, with centralisation and with the technical and social instability inherent in trying to get so many ergs from one basket as large as a 1300 to 2000 MW reactor or even a 10 000 MW nuclear park. Smaller units are inherently more stable in the broadest sense. One is more likely to get public acceptance of them. They are more sensible on energy grounds because small units can be sited close to urban and industrial centres, which can also use the waste heat from the power stations. If you have low electricity growth, as we are forecasting, you may run into serious trouble if you start putting on-stream new units in large chunks of 2000 MW or so. Much more stable in terms of the power station construction and electricity equipment industries is to have a series of small orderings rather than occasional large units every few years. There is, in many countries, a rapidly growing perception that material prosperity can be coupled to zero energy growth or even to declines in those commodities. In Britain, our abundant indigenous fuels give us the gift of time in which to debate and to decide with fuller information than we have had during this meeting whether or not we should take the fast breeder route, or follow a large range of alternatives. There is no great urgency to make irrevocable commitments.

REFERENCE

1. Leach, G., Lewis, C., van Buren, A., Romig, F. and Foley, G., *A Low Energy Strategy for the United Kingdom,* Science Reviews Ltd, London (1979)

5

Technology and Energy Supply

Meredith W. Thring

SUMMARY

In order to avoid World War III or other world-wide disaster, it is essential to have a fuel policy in which the developed countries bring their per capita energy consumption down to about the present world average figure (1.8 tce per capita per annum) over the next 30 years (3–4% per annum reduction). The use of the premium fuels, electricity, natural gas and oil, will have to be reduced at about twice this rate if the underdeveloped countries are to have a reasonable share; this is essential to reduce world tensions. Thus, in particular, Britain needs no new power stations, but will have to convert existing ones to the use of pass-out steam.

5.1 ENERGY STRATEGY

Thring's Principle of Economics states, 'Whatever is right for our grandchildren is always uneconomic now and almost always impolitic.'

Nowhere is this principle more obviously true than in the field of energy strategy. It is apparently economic and politic to do the following in Britain.

(1) We plan a perpetually increasing total energy consumption per capita, although we are already using three times the world average figure (1.8 tce per capita per annum). The only argument is about whether it should increase at 4% per annum or only 2%. Less growth is regarded as an impossible ideal and actual decrease as nonsensical, although we already know how to reduce the figure by 30–50% with no loss of output or convenience merely by investing in known fuel-saving devices.

(2) We use electricity for space heating and water heating when the total capital cost is several times as great as the direct fuel heating system and the energy consumption is nearly four times as high. We are in fact lagging far behind all other developed countries in the use of total energy systems (electricity generation, waste heat utilisation).

49

(3) We have reduced our coal output since the war from about 200 million tonnes per annum to less than 120 million tonnes per annum and this energy gap has been entirely filled with oil. It is known that world oil reserves will certainly begin to become much less accessible in less than 40 years, whereas the coal reserves are sufficient for 200–300 years.

(4) We have reduced our rail network and encouraged road transport when the railway is the only system that can run as conveniently on coal as on oil.

(5) Cars and lorries are sold on the basis of power, acceleration and top speed and no serious attention is given to fuel economy.

(6) We are steadily converting agricultural land into motorways.

In the next sections of this paper, I shall try to show that the true interests of our grandchildren who will be living in the twenty-first century require the following energy strategy.

(1) A reduction in the energy consumption per capita in the developed countries to reach about the present world average figure in about 30 years time. This must be associated with an increase in the energy consumption per capita in the less developed countries to the same figure. Since the rate of population growth in the less developed countries is 3% per annum and in the developed countries is 1%, the world will have doubled in population in about 30 years. With the same average energy consumption figure as at present, the world total energy consumption will double on this strategy. This energy consumption will be

$$1.8 \times 8000 \times 10^6 = 1.44 \times 10^{10} \text{ tce per annum.}$$

For Britain, this policy requires an annual decrease of total energy consumption per capita between 3 and 4%.

In the third section of this paper, I shall show how painlessly this can be achieved.

(2) The premium fuels, oil, natural gas and electricity, must be reduced at a rate about twice as fast, and the difference must be taken up by increasing the use of the non-premium fuels, coal, waste heat, refuse and renewable (solar, wind, etc.). Oil and natural gas are premium fuels because there are not enough reserves to give all mankind a fair share of them and they are so ideally suited for certain purposes (transport, petrochemicals, small burners) that we shall have to synthesise them at vast cost when they are exhausted. Electricity is a premium fuel because the capital cost of the large central generator and the distribution system is so high that it could not possibly be provided for all 8000 million people; so it is a 'rich man's toy' except for small local generators.

It follows that the whole expenditure on nuclear reactors is an evolution like that of the Brontosaurus in a totally unviable direction. Our policy is based on a fundamental fallacy—that we need more electricity. We are already long past the optimum saturation value.

(3) We should divest all the vast sums at present spent on research and development of nuclear power to the following:

(i) Developing a method of getting all coal out of the ground without men ever going underground (telechiric mining).

(ii) Applying total energy systems to existing coal-fired power stations, including those that are regarded as inefficient.

(iii) Using fiscal means (e.g. a tariff on premium fuels spent entirely on low-interest loans for fuel-saving equipment) to make it economic to save fuel by known methods. This applies to vehicles, domestic heating (especially new building) and industry.

(iv) Research on and fiscal encouragement of the small-scale local use of renewable energy resources (solar, wind, water power, CH_4 from agricultural and city refuse, refuse combustion).

5.2 THE INTERLINKAGE OF ENERGY AND OTHER PROBLEMS

As soon as one looks at the energy problem on a long-term strategic basis, one finds that it is so closely connected to three other groups of problems that it is essential to develop a strategy to solve all of them simultaneously.

(i) It is certain that if the gap in standard of living and energy usage between the developed and non-developed countries continues to grow, and the population of the latter grows at 3% per annum while that of the former only grows at 1% per annum, then the resulting world tensions will inevitably escalate us into World War III. Already the war areas of the world are all ones where there is a sharp local contrast between rich and poor groups distinguished by religion, race or nationality. The Stockholm International Peace Research Institute has

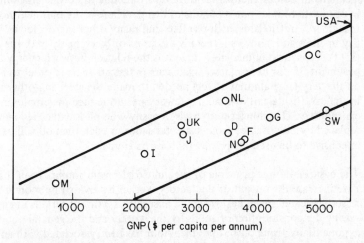

GNP ($ per capita per annum)

Figure 5.1 Relationship between energy consumption per capita and gross national product for 12 countries in 1972. Key: USA, United States of America; C, Canada; SW, Sweden; G, Germany; NL, Netherlands; D, Denmark; F, France; N, Norway; UK, United Kingdom; J, Japan; I, India; M, Malaya.

51

estimated that 40% of the world's scientists and engineers are working for war purposes either directly or indirectly (e.g. by rocket or Pu research). It is universally agreed that the use of a fraction of the weapons already available means the inevitable end of our civilisation.

Figure 5.1 shows the relation between the standard of various countries and their energy consumption expressed in tons of coal equivalent per capita per annum (1 tce = 2.4×10^7 kJ). The relation is nearly linear except that those developed countries such as Britain, which had traditionally had very cheap indigenous fuel resources, come on the upper limit of the band, i.e. they get less financial benefit out of their energy consumption than the average. The average for all the developed countries is between 5 and 6 tce per capita per annum while that for the other countries is 0.5 tce per capita per annum and the world overall average is about 1.8 tce per capita per annum. The energy equivalent of a full diet is about 0.19 tce per capita per annum (3000 kcal per day = 1.25×10^4 kJ per day = 4.58×10^6 kJ per annum), so that the average person in the developed countries has nearly 30 energy slaves working for him, while in the others he has less than three.

(ii) There are a whole group of consequences to human physical and mental health from our careless use of energy in the developed countries; these can be classified under the general heading of lowering quality of life by pollution. There is the chemical pollution of air, land and water by, for example, Pb from motor car exhausts, SO_2 and SO_3 from combustion of sulphur-containing fuels, CO, soot and unburnt hydrocarbons from bad combustion, heavy metal and cyanide discharge or dumping from factories, and radioactive escapes. There is the inhalation of dangerous chemicals by workers in factories and mines; many accidents in homes, factories and transport are due to cutting economic corners in our haste to raise our standards of living. (There is the prevalent noise of machinery, traffic jams, high-rise flats and many other factors lowering life quality in cities and, above all, there is the stress of 'keeping up with the Joneses'.)

(iii) Perhaps most immediately urgent is the relation between energy and unemployment. If, for example, we make cars to last 50 years in order to economise on the very large amount of fuel needed to make the steel and other materials in the car, then clearly we sell far fewer cars and reduce the employment in the car industry. On a longer-term basis, if easily won oil is exhausted and has to be replaced by oil obtained from shale, tar sands or coal, then oil will be too expensive to be used for private motor cars anyway.

The basic economic principle of the industrial revolution has been that, whenever we increase the output of the factory worker by giving him more power and more sophisticated machines, we could absorb this without serious unemployment by raising the consumption of goods by the worker and the general public. Thus, an exponentially rising output per man hour has been associated with an exponentially rising consumption of energy and raw materials. Like all exponentially rising curves, the consumption curve has reached its limit or very close to it, due to limitations in conveniently won raw materials, in the unfulfilled needs of the underdeveloped countries, limitations of space (e.g. on city roads and in homes and kitchens) and the increasing resistance of the public to gadgetry and fashion changes. Thus, unemployment is rising inexorably in all the developed countries, while in the underdeveloped ones the population growth rate is causing a drift into workless shanty towns.

Thus, the present system of gearing the economic system to an exponentially rising energy consumption will lead to a total breakdown of the economic system if we cannot change it in time.

The recent survey by the Workshop on Alternative Energy Strategies* assumed 3-5% per annum economic growth in 15 non-Communist major oil-importing countries (i.e. developed countries)—a reduction from the 6% per annum growth in 1960-1972. They concluded that, even without any growth in the use of oil by the less developed countries, the supply of oil will fail to meet increasing demand before the year 2000, and that the developed countries will have to shift from oil to other energy sources in 5-20 years. These industrial countries are at present obtaining more than half their total energy from oil.

Two other energy facts must be pointed out. In the first place, within our present knowledge there is no substitute for liquid petroleum for air, sea or road transport which would have comparable availability and convenience. This is because it is a liquid of high calorific value which can be carried in light tanks and pumped in small tubes and because it burns with some 14 times its own weight of air that is picked up as the vehicle goes along. The only comparable system would be a fuel cell consuming a cheap liquid fuel such as methanol with air as the other reagent to make electricity. Such a system could be developed, but has not yet been. Similarly, agriculture in the developed countries is entirely dependent on oil for its high output, both for tractors and for firing nitrogen and making pesticides and sprays.

Liquid petroleum also constitutes an excellent and abundant feedstock for the whole petrochemical industry for which there is no satisfactory substitute. It follows that if we, in the developed countries, use up all the readily available petroleum by (1) wastefully overpowered vehicles, (2) using it for heating and (3) making throwaway plastic goods, purposes for which coal is perfectly satisfactory, we are condemning our grandchildren to a very much less convenient life, while we are making it impossible for the less developed countries ever to have the benefits of our kind of technology.

Finally, we have the practical consequence of the laws of thermodynamics that a thermal power station, whether fossil-fuel or nuclear fired, which does not have pass-out steam or other waste heat utilisation (of all British power stations, only Battersea has this), has an overall thermal efficiency from potential energy to electricity supplied to the user of barely 25%. This means that three-quarters of the heat produced by the reaction is lost to the atmosphere in chimneys, cooling towers, mechanical friction and Joule heating of generator coils, transformers and distribution wiring. By installing a pass-out steam heating system, over 70% of the reaction energy can be used.

*See: Flower, A. R., 'World Oil Production', *Scientific American*, March 1978, p. 42.

5.3 ENERGY IN THE LESS DEVELOPED COUNTRIES

Even from a purely national point of view, our grandchildren will have no future in the next century if we continue to fail to relieve the poverty of the less developed countries by giving them the tools to help themselves. (This follows because (1) no country will escape the consequences of World War III, (2) pestilences will inevitably spread as have 'flu epidemics, and (3) international callousness leads to local callousness.) This leads to the following:

Conclusion I: The long-term best interests of any country are identical with those of all mankind.

From this immediately follows conclusion II.

Conclusion II: We must find a humane way of levelling off the world's population in one generation.

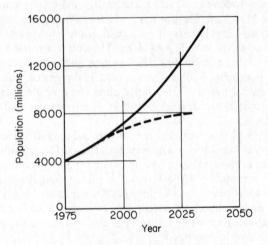

Figure 5.2 World population growth: ———— 2.25% per annum continuous growth; —————— desirable levelling off.

While we can support 8000 million people on the Earth (twice the present number) with the available energy and agricultural land, it is surely impossible to support 16000 million by any feasible development of technology within the resource constraints (see *Figure 5.2*). The only humane way of doing this is to reduce the population growth rate of the less developed countries to that of the developed countries by providing a decent standard of living and a decent education for all humanity.

Only the rich countries have the spare resources of capital and skilled manpower to do this and thus the future of civilisation lies in their hands.

It has been stated above that those who have studied world peace have concluded that this is impossible as long as the world is in a state of tension due to the gap between standards of living in rich and poor countries. This gives us our next conclusion.

Conclusion III: To avoid World War III, we must essentially eliminate the gap in standards of living between developed and underdeveloped countries within 30–40 years.

This gap is well expressed by the energy consumptions averaging 5.5 tce per capita per annum in rich and 0.5 tce per capita per annum in poor countries.

Figure 5.3 shows various possible scenarios for the future use of energy per capita in the two groups of countries. If the developed countries insist on increas-

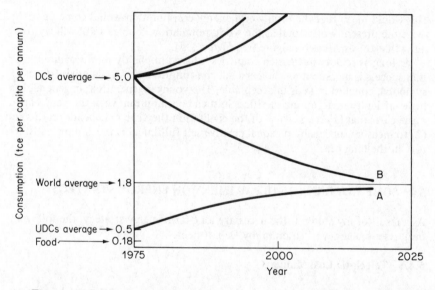

Figure 5.3 Present and future energy consumption per capita per annum.

ing their energy use by 2–4% per annum, then the world consumption of energy will be more than six times the present figure when the world population doubles in 30 years time. This is not feasible because of the points now outlined.

(i) Oil and natural gas from direct sources will be completely exhausted, and coal, tar sands and oil shale will have to be used at about 20 times the present rate while the energy needed to build the conversion equipment will be impossibly high.

(ii) This amount of energy could not possibly be provided by nuclear fission, even breeders, because of the capital cost of building 1000 MW stations and the associated distribution systems all over the world to give 8000 million people the energy equivalent of six tons of coal each. Moreover, as such stations last only about 30 years, the world would become steadily littered with derelict concrete erections which could never be dismantled.

(iii) Thermal pollution of lakes and rivers, steam from cooling towers and the CO_2 content of the atmosphere would all rise to dangerous levels. For example, the greenhouse effect could cause a significant rise of sea level due to thawing of the polar icecaps.

Thus we are forced to arrive at our final conclusion.

Conclusion IV: The only feasible equilibrium scenario for the twenty-first century is one in which the rich countries come down to an average energy consumption about equal to the present world average (1.8 tce per capita per annum) and the poor countries come up to a similar figure.

This would mean that the total world energy consumption would come up to twice the present world total as the world population doubles and I will try to show that this is a feasible figure (see *Figure 5.3*).

As long as society in the rich countries is based implicitly on the assumption that success is measured by the personal consumption of raw materials (status symbols), conclusion IV is unacceptable. The younger generation, in general, has little difficulty in accepting sacrifices in such consumption when the points are made clear that (1) the survival of the civilisation they live in depends on this and (2) to measure success by personal creative self-fulfilment is much more satisfactory in the long run.

5.4 SOME WORK IN THE DEPARTMENT ON ENERGY STRATEGY

As a result of my study of the necessary long-term energy strategy, the following projects are being worked on in my Department.

5.4.1 Telechiric Coal Mining

This means that the miner does the same job as at present, cutting and transporting coal with the same machines, but he does all the tasks from the surface by controlling telechirs, which are in such good communication with him that he can use all his trained craft skill as though he were down the mine. The telechirs consist of electrically powered bodies which can move in the mine wherever a man could move, carrying hands with force feedback and visual communication systems. The advantages of this system are given below.

(i) The miner is not subject to danger and a daily journey to the face, and does not run the risk of contracting pneumoconiosis.
(ii) The process is much cheaper, because it is not necessary to ventilate the mine, to have fire precautions nor a human travel and life support system.
(iii) It is possible to win coal from thin seams, very deep seams and seams far under the sea.

56

5.4.2 Hybrid Diesel/Electric Passenger Car

We held a Conference on Fuel Economy in Vehicles at Queen Mary College in 1972 (before the oil crisis) and it emerged that the hybrid diesel/electric was the most feasible way of halving the fuel consumption of a small passenger car. We are building one with existing components and hope to have it running in early 1979.

5.4.3 Coal-Fired Steam Locomotive

We are making a study on a coal-fired steam locomotive of high efficiency and it appears that we should be able to obtain an overall efficiency better than a diesel electric.

5.4.4 Steam Engine

We are starting to develop a small steam engine for the hotter countries which will be solar heated in the day and fired with local waste materials at night.

5.4.5 Walking Tractor

We have built a small model of a very simple walking tractor which will use much less power than wheels for the same traction at low speeds.

Part II
Risk and Uncertainty

6

Fast Reactors and Problems in Their Development

Norman Dombey

I want to discuss here the main differences between fast reactors, in particular the liquid-metal fast breeder reactor (LMFBR), and thermal reactors. As a physicist, I take the view, based on the intrinsic physics of the systems, that fast reactors should be considered as a different genus from thermal reactors. I hope that a discussion from this viewpoint will illustrate some of the problems of fast reactor development. I should also like to introduce to non-specialists some of the features characteristic of fast reactors, as this information is not readily available outside the technical literature. Finally, I shall draw some conclusions for fast reactor development generally and for the British programme in particular. These conclusions will be almost identical to those reached by John Surrey[1] even though he approached the problem as an economist, not as a physicist.

I will not have space, unfortunately, to say anything about the implications of the widespread use of plutonium, although I am a member of the British Pugwash Group and share the concern of the International Pugwash Council over the dangers of proliferation of nuclear weapons inherent in an early commitment to fast reactors[2].

We start, then, with the physical differences between thermal and fast reactors. The main points are given in *Table 6.1*.

The most important point is the last: thermal reactors are designed to be in their most reactive nuclear configuration. This follows from the presence of a moderator, without which the natural slightly enriched uranium fuel of the reactor would not go critical at all. The corollary of this for fast reactors is that, as there is more than one critical mass of fissile material in the core of a fast reactor and no moderator, it is not difficult to envisage a rearrangement of fuel to provide a more reactive configuration. As the reaction goes by fast neutrons, it is possible to attain a prompt critical reaction with a large energy release over a very short time. (About 0.2% of the neutrons in a fast reactor are delayed, i.e. come from radioactive daughter nuclei with various half-lives of the order of one second: it is these delayed neutrons that allow reasonable control of the fast reactor because the prompt (direct) neutrons are captured in about one microsecond (10^{-6} s). A fast reactor is therefore designed to be just critical including the delayed neutrons, but

61

TABLE 6.1 THE PHYSICAL DIFFERENCES BETWEEN THERMAL
AND FAST REACTORS

Thermal	Fast
Natural or slightly enriched (2–3%)	20–30% enriched uranium or Pu fuel
Less than one critical mass	More than one critical mass
Moderator	No moderator
Thermal neutrons ($\sim 1/40$ eV)	Fast neutrons (keV–MeV)
Most reactive configuration for nuclear reaction	Not most reactive configuration for nuclear reaction

not critical without them. A prompt critical reaction is one which is critical by prompt neutrons alone.) This is called a nuclear excursion and cannot occur in any of the normal types of power-generating thermal reactors.

Another important point in this comparison follows from the absence of a moderator in the fast reactor which allows the fast reactor core to be much smaller than a thermal reactor of the same power. So the power density of a fast reactor is about five times that of a light-water reactor (LWR). (We should remember that LWRs themselves have a power density about ten times higher than British gas-cooled reactors.) So the coolant used in a fast reactor must have very high thermal conductivity and all the present prototypes use liquid sodium.

Richard Wilson[3] has written a review article on the physics of fast reactors which is a very useful reference on the subject. He has his own table on the important differences between LWRs and LMFBRs which is reproduced as my *Table 6.2*.

The coolant in a LWR is water, which is also the moderator. Wilson's second and third points again illustrate my last point in *Table 6.1* that the core configura-

TABLE 6.2 IMPORTANT DIFFERENCES BETWEEN LWRs AND LMFBRs
(from a review[3] by Richard Wilson)

LWR	LMFBR
Operates with coolant under pressure	Operates with coolant under atmospheric pressure
Removal of coolant shuts down reactor	Removal of coolant may increase reactivity
Compaction of core decreases reactivity	Compaction of core increases reactivity
Power density ~ 80 W cm^{-3}	Power density ~ 400 W cm^{-3}
Post-accident heat removal needs emergency cooling system	Post-accident heat removal from surrounding sodium

tion in normal use of an LWR is of maximal reactivity, whereas this is not the case for a fast reactor. Nevertheless, the LMFBR, unlike an LWR, does not need a thick-steel pressure vessel and, although there are serious difficulties in the use of sodium as coolant, its thermal capacity is so high that no emergency cooling system is needed in the design of an LMFBR.

It is for the job of the licensing authorities in each country to consider hypothetical accidents, to insist on measures to minimise the risk of accidents and to minimise and contain their effects if they do occur. The discussion of hypothetical accidents begins with an analysis of the reactivity coefficients of the reactor. These coefficients measure the change in reactivity of a reactor as the physical state of the reactor changes. The important coefficients for safety purposes are those related to an increase in power and temperature of the reactor: if the power or temperature increases, a negative coefficient implies that the reactivity will decrease. As commercial thermal reactors operate in the configuration of maximal reactivity, all nuclear reactivity coefficients in these thermal reactors will be negative. For fast reactors, the reactivity coefficient which allows these reactors to be considered as potentially suitable for commercial use is the Doppler coefficient. This measures the change of neutron reaction rates with temperature in the reactor materials themselves. As the temperature rises, the average neutron energy increases; for uranium-238, neutron absorption increases strongly with neutron energy and, hence, there will be fewer neutrons available for fission, so the Doppler coefficient is large and negative.

The fissile material in the core will be plutonium or ^{235}U: for ^{235}U the Doppler coefficient is positive in the same temperature range, whereas for plutonium it is small[4]. So for a fast reactor of the design envisaged in CFR-1 or Superphenix, with about 20–25% plutonium and 75–80% natural uranium (99.3% ^{238}U) in the fuel rods, there will be an overall negative Doppler coefficient.

It is important to emphasise that the Doppler coefficient, although dependent on core geometry, is primarily an intrinsic nuclear quantity. It depends on the nuclear composition of the fuel used and its arrangement in the core; once an acceptable configuration is found, there is no reason why this configuration should not be scaled up to a larger reactor design. The Doppler coefficient is the most important intrinsic safety mechanism of the present generation of fast reactors.

The second most important reactivity coefficient in fast reactors is the sodium void coefficient. Although fast reactors do not have a moderator, the sodium coolant flows in the core and scatters the neutrons, so to some extent it acts like a moderator in slowing down the neutrons. It also absorbs some neutrons.

If the core were to overheat, the sodium could begin to boil; the void coefficient then measures the change in reactivity of the core caused by bubbles of sodium gas forming in the liquid sodium flowing through the core.

Here, there are two competing effects. First, sodium gas would absorb and scatter fewer neutrons than would liquid sodium as it is less dense, so there would be more neutrons of higher energy around and the reactivity would tend to increase. Secondly, the presence of sodium bubbles would increase the probability of neutrons escaping from the core through the bubbles, so this would tend to decrease the reactivity.

It is easy to see that the first effect, which leads to a positive void coefficient,

63

is proportional to the volume of sodium in the core whereas the second effect is an edge effect, which is proportional to the effective surface area of the sodium. So for small reactors, where all the sodium can be taken to be near the surface, the edge effect wins and the void coefficient is negative. This was the case for the old 15 MWe Dounreay Fast Reactor. As the reactor gets larger, the volume effect will begin to dominate and the present prototypes such as the Prototype Fast Reactor (PFR) or Phenix do have a positive void coefficient, albeit small. The planned CFR-1 or Superphenix, however, will have a large positive void coefficient, about five times larger than that of the present prototypes.

So scaling up from a 250 MWe prototype to a 1300 MWe 'commercial' reactor is not simple: the problems connected with sodium voiding get much worse.

There are other reactivity coefficients resulting from the expansion of the fuel rods and melting of the cladding surrounding the rods, but these are not of primary interest here.

In addition to a comparison of reactivity coefficients, three other areas need to be considered in any comparison of fast reactors with thermal reactors: the fuel–coolant interaction, the other non-nuclear problems arising from the use of liquid sodium as coolant, and the general problem of recriticality.

The fuel–coolant interaction is not a nuclear effect at all: it is the potentially explosive interaction arising when a molten metal comes into contact with a liquid at a temperature much above the boiling point of the liquid. It is well known[5] in the steel industry from incidents where molten steel has come into contact with water. It is a problem inherent also in water-cooled thermal reactors, as it has been known for a piece of hot fuel rod to melt and interact explosively with the surrounding water. However, it has been pointed out[6] that '. . . the safety concerns for the two systems (LWRs and LMFBRs) are somewhat different. With respect to LMFBRs, the questions of vapour explosion and milder molten fuel–coolant inter-actions leading to coolant voiding (with a potentially positive void coefficient) are both of primary importance. For LWRs, however, the effects of coolant expulsion are not as critical in comparison, due to an overall negative void coefficient. . .'.

The best known difference between thermal reactors and the LMFBR stems from the difficulties of using liquid sodium as coolant in the LMFBR. Sodium is, of course, highly corrosive and interacts vigorously with both air and water on contact. All those large prototypes at present operating (PFR, Phenix and the Soviet BN-350) have been plagued by sodium leaks. This is a purely non-nuclear engineering problem, but the use of sodium may well represent a basic design flaw in present fast reactor technology.

An official US report[7] on problems of LMFBR steam systems puts it thus: 'The sodium-heated steam generator system is one of the most critical of the non-nuclear elements of an LMFBR plant, owing to the demand on it for extremely high reliability of the sodium to steam/water boundary and the necessity to make the system capable of safely accommodating any failure of this boundary and the resulting sodium/water reaction that could occur--all to be accomplished with assured functional performance and tolerable equipment and operational costs. Despite the intensive steam generator development in the LMFBR community throughout the world since the 1950s, the assurance that the designs now being implemented will in fact be adequately reliable for commercial plants is clouded by the difficulties with leaks, tube vibration and flow instabilities that prevented sustained operation of the United States Fermi plant steam generators in the 1960s, the numerous small leaks that have delayed start up of the British PFR since late 1974 and the large leak sodium/water reaction incidents that have occurred in the Russian BN-350 beginning in 1973.'

Since that report was published, several sodium leaks caused Phenix to be completely closed down between October 1976 and June 1977.

In fast reactors, there are also recriticality effects to consider. As the core is not in its most reactive configuration and there may be over hundreds of critical masses of plutonium in it, a disruption of the core may lead to a much more reactive configuration: for example, a coolant channel gets blocked; the sodium boils; the reactivity increases because the void coefficient is positive; the cladding of some fuel rods melts and pieces of fuel containing plutonium fall and form a molten mass of plutonium which, if large enough, can then become prompt critical leading to a rapid release of energy. Or there is a scenario involving the fuel–coolant interaction whereby a piece of hot uranium–plutonium fuel breaks off and comes into contact with the sodium coolant well above the boiling point of sodium; it is then injected explosively into another fuel rod containing plutonium. Under such implosive conditions, a sub-critical piece of plutonium can become prompt critical leading again to a rapid release of energy.

The analysis of the energy release and the pressures created by such hypothetical core-disruptive accidents is being pursued in all the countries engaged in fast reactor technology. Enormously complicated computer programmes are being prepared to simulate these conditions. Whether or not the present computer programmes are reasonably accurate in describing fast reactor accidents, it is clear that the role of negative reactivity coefficients will be to limit the energy release whereas positive coefficients will enhance it. Thus, for a reactor of the size of PFR or Phenix with uranium–plutonium fuel, the energy release for a core-disruptive accident of the type described is not expected to be larger than 400 MJ (equivalent to about 200 lb of TNT). But in a reactor the size of CFR-1 or Superphenix, the sodium void coefficient is much larger and positive, so the energy release for a similar accident could be 2000 MJ (equivalent to 1000 lb of TNT).

Now the point of this discussion is not that fast reactors, or large fast reactors, will blow up. It is the job of the licensing authorities, after all, to ensure the safety of such systems. The point of the discussion here follows from the fact that the licensing authorities will insist

(*a*) on minimising the risk of an accident,
(*b*) on limiting the damage to the reactor core in the event of an accident,
(*c*) on maintaining the integrity of the primary reactor system even in the event of a core-disruptive accident leading to a large energy release, and
(*d*) even in the worst case of an accident breaking the primary system, that the radioactive materials released are contained.

These are the four lines of assurance of the US authorities, but the practice of the British Nuclear Installations Inspectorate should be similar[8].

The point that I want to make is that the licensing authorities will insist on measures to deal with the safety and other engineering problems specific to fast reactors, and that these measures will add appreciably to the capital costs of fast

TABLE 6.3 COST ESTIMATES OF THE 300 MWe LIQUID-METAL FAST BREEDER PROTOTYPE PLANT (SNR-300) DEVELOPED IN WEST GERMANY (in million DM, at constant prices)

Number and date of estimate	Price index (July 1965 = 100)	Source of estimate	Plant excl. owner's costs and first core	Owner's costs	First core			Plant incl. first core	
					Fabrication	Fuel	Total	Excl. fuel and owner's costs	Incl. fuel and owner's costs
1 July 1965	100.0	Karlsruhe							300
2 Oct 1965	100.3	Karlsruhe	259						309
3 Dec 1965	100.5	Ministry/Atom Commission					50		448
4 May 1967	102.3	Karlsruhe and Interatom	254						
5 8 Dec 1969	116.7	Konsortium SNR	471						
6 31 Dec 1969	116.7	Konsortium SNR	443						
7 1 July 1970	122.2	Konsortium SNR	505						
8 11 Feb 1971	131.2	Konsortium SNR	511		46	38		557	
9 15 April 1971 (first tender)	132.0	Konsortium SNR	548	69	45			593	664
10 1 Oct 1971	133.8	Konsortium SNR	704		39			743	
11 7 Feb 1972 (binding tender)	137.0	Konsortium SNR						859	
12 10 Nov 1972 Supply contracts	138.4	SBK/INB	679						
Provisions for additional costs			168						
Total			847	74	43	36	79	890	1012
13 Oct 1975	138.4	SBK/INB	1040	102	43	38	81	1083	1223

reactors compared with thermal reactors of a similar size. Yet the goal of a commercial fast reactor system is to produce cheaper electricity than other systems such as thermal reactors. It has been demonstrated[9] that fuel costs play a minor role in the total cost of nuclear electricity, so that fast reactors will only be competitive with thermal reactors if their capital costs are comparable, unless the price of uranium were to increase dramatically. The special safety requirements of fast reactors and the difficulties of using sodium in LMFBRs both suggest that the capital cost of an LMFBR will be much higher than a comparable thermal reactor. Furthermore, as has been indicated, the safety problems get worse for large fast reactors. It may be thought that my comments are overly pessimistic or overly theoretical—I am, after all, a theoretical physicist. But the one case study available on the capital costs of fast reactors does indeed bear out my conclusions.

Otto Keck has written a thesis[10] on the prototype German fast reactor SNR-300. *Table 6.3*, taken from his thesis, illustrates the changes in costs (at constant 1972 prices) between the initial estimate of 300 million DM at the time the reactor was authorised to the most recent estimated cost of 1223 million DM of October 1975. These are not the final figures and the cost escalation is therefore by at least a factor of four in real terms. I should say also that Keck demonstrates that the initial cost estimate of the SNR-300 was in no way based on any evidence about the likely capital costs of fast reactor systems. The estimate was simply a number which the fact reactor proponents thought to be reasonable in view of the likely cost of a thermal reactor system of comparable size. I suspect that the figures[11] we have heard from the AEA about the likely cost of CFR-1 have been arrived at in a similar way. Keck[12] now estimates that the capital cost of the SNR-300 is five times that of a comparable LWR system ordered at the same time.

The importance of this case study is that the SNR-300 is the first, and so far the only, prototype fast reactor that has been built as if it were a commercial reactor: that is to say, it has been built specifically with the goal of satisfying both the utilities and the licensing authorities. Therefore, the utilities and the licensing authorities have been consulted from the early design stage in 1969.

For the results of this consultation, it is best to use Keck's own words from Chapter 6 of his thesis[10]:

The impact of licensing requirements on plant design and construction costs
After the designs were submitted at the end of 1969, drastic alterations became necessary as a result of the safety and environmental provisions demanded by the licensing authorities. These went far beyond those incorporated in the design as submitted at the end of 1969 and substantially increased the costs of the plant. Together with other factors, the licensing requirements delayed the start of construction by about two years.

The most severe licensing requirement related to hypothetical accidents. The safety philosophy of the original design placed the emphasis on measures to prevent a core meltdown and a subsequent hypothetical nuclear excursion, and it was thought that this accident could be made sufficiently unlikely to avoid measures accommodating its consequences. The licensing authorities, however, not only required more stringent measures for prevention of this accident, but also stipulated that the impact of a nuclear excursion should be safely accommodated by the reactor vessel and the containment. It was specified that the reactor vessel should be designed to accommodate a power burst up to 150 MW s. The hypothetical case that the reactor vessel would not remain intact

67

was to be provided for by designing the containment housing the primary coolant circuit to withstand a power burst up to 370 MW s. In addition, the hypothetical case of a collapsed reactor core melting through the reactor vessel, and breaking through the structures beneath the vessel into the ground, was to be guarded against by installing a 'core-catcher', a device which can safely receive molten fuel, prevent it from forming a critical configuration and cool if sufficiently to avoid damage to the containment.

A redesign of the primary coolant circuit was required to guard against the hypothetical case of loss of coolant through breakage of one of the main coolant pipes. The components of the primary coolant circuit such as the reactor vessel, pumps, and intermediate heat exchangers had to be arranged in a way so that a minimum sodium level is maintained whatever breakages in the piping or the components occur. With this minimum sodium level, the decay heat from the core can be removed, either by a coolant loop, if one remains intact, or by an emergency cooling system. In order to eliminate possible sodium voids through gas bubbles entering the reactor core, a special device had to be designed in the coolant inlet of the reactor vessel which separates the gas from the sodium before the coolant passes through the core.

Altogether, the increase in the construction costs resulting from the stipulations of the licensing authorities was enormous. A rough illustration of the impact of licensing requirements on plant design and plant costs can be given by comparing the size of the reactor vessel in the first design of December 1969 and of the revised design of 1971. The costs of the design changes already included in the contract of November 1972 can be estimated only with a high margin of uncertainty, as this estimate cannot be based on a comparison of binding bids. Subjective estimates place these costs at 60–80 million DM, out of a total amount of 1000 million DM, as fixed in November 1972 in the supply contracts. Later design alterations, resulting from licensing requirements, for which exact cost figures are available, caused until October 1975 a cost increase of 366 million DM (in constant 1972 money). It can be expected that further cost additions will be incurred as the licensing procedure goes on. The whole impact of licensing requirements on plant costs will be known only after the last part of the licence is issued and construction is completed.

The demands of the licensing authorities not only increased the costs of the prototype plant, but also necessitated a lot of additional research and development in industry and government laboratories, which added considerably to the total costs of the fast breeder project. For some items, the licensing requirements even went to the limits of technical feasibility, and imposed heavy tasks on scientists and engineers in industry and government laboratories.

So, on the basis of this case study, there is every reason to believe that the licensing requirements for the fast reactor, arising specifically from the physics problems I have outlined, will cause an appreciable increase in capital costs relative to thermal reactor systems. Furthermore, the situation will get worse for larger fast reactors such as CFR-1 and Superphenix.

As Bupp and Derian[9] showed, the relative higher capital costs of fast reactors compared with thermal reactors must be weighed against the prospective future price of uranium. I hope that I have demonstrated that the likely capital cost of a fast reactor system will be appreciably more than that of a thermal reactor of similar size, yet although, when Bupp and Derian wrote their article in 1974, future uranium supplies looked uncertain and the US authorities were investigating using low-grade and therefore expensive uranium from the vast shale deposits in the United States, uranium supplies for the rest of this century from high-grade suppliers are now assured.

In the words of the Times Mining Correspondent[13] 'colossal new discoveries of uranium' have been made in the past three years in Australia and Canada. Therefore 'prices are unlikely to go up faster from their current $42.3 per pound than inflation warrants'. So there can be no urgency about developing a commercial fast reactor for economic reasons: it would be more cost effective to spend money on improving the present designs of thermal reactors.

Although the AEA talk about a commercial fast reactor, there is no sense at present in using the term commercial. The present state of the art of fast reactor safety-related research in the United States is reproduced in *Table 6.4* from Wilson's review article[3]. It shows that there still remains an enormous amount of basic work to be done on these systems before they can be regarded as a safe, sure and reasonably priced source of electricity. Note also the part about 'commercial' reactor systems in section (4)E. This is due, of course, to the large size of the sodium void coefficient, as has already been explained.

My final point is that fast reactors are too hard for us in Britain working alone. We have neither the resources in money or the resources in people to do the job by ourselves. The greatest part of the AEA's budget and most of its qualified scientists and engineers since the mid-1960s have been committed to the fast reactor. Nevertheless, we cannot match the spending and manpower committed in the United States on fast reactor systems, even though they do not even have an operating reactor.

Sir Brian Flowers made the same point some years ago[14]: 'But I see no need for undue speed, even if one believes there is no avoiding fast reactors and the plutonium economy they entail. Thermal reactor development has proved to be within the capacity of the United Kingdom, but only just. Fast reactor development is likely to be beyond our resources. It seems to me essential that if we pursue it at all we should do so on a European basis.'

I would fully support this view. Britain is favourably placed for energy until the end of the century. Fast reactors cannot hope to be commercial until next century. A modest programme of research and development is indicated for Britain, which does not strain its limited resources, so that it is in a position to benefit from any future exploitation of fast reactors next century. The problem is similar to that in the field of fusion power and the solution is similar: Britain should join in a common European programme of research and development, similar to the Joint European Torus (JET) and similar to the programme of research in elementary particle physics based at CERN in Geneva. If it is too late for Britain to join in the Superphenix project, it should indicate its willingness to join in the next reactor after that. With its continuing experience of PFR at Dounreay and its large and assured stock of plutonium, it should be assured of a welcome as a partner.

So it does not make sense for Britain to build CFR-1. On the other hand, there is scope in Britain being involved in planning the reactor to be built as a European collaboration (perhaps even the US and Japan would also be interested in participating) after Superphenix. The considerations I have outlined here would suggest that a sensible size for a LMFBR which avoids the problems connected with a

69

TABLE 6.4 PRESENT STATE OF THE ART OF FAST REACTOR SAFETY-RELATED RESEARCH IN THE US—KEY ISSUE AND REQUIRED ACTIVITIES MATRIX (from a review[3] by Richard Wilson)

Key safety issues and problems	A Brief description	B Impact on design/ licensing	C Current status	D Major remaining uncertainties	E Summary/probable resolution of issue
Pin to pin propagation (1)	What are the limits to which we can operate long-term with failed fuel in the reactor?	Is individual sub-assembly instrumentation required?	Operation with limited failed fuel appears feasible.	Long-term operation with failed fuel needs to be demonstrated.	Long-term operation with failed fuel likely to be demonstrated.
	Can local faults (blockages, etc.) propagate, and if so, what are the limits of detectability?	Are whole core sensors required?	Rapid pin to pin propagation appears unlikely.	Slow blockage propagation cannot yet be ruled out.	If blockage propagation should occur at all, it can be detected by appropriate instrumentation.
		If pin to pin propagation does not occur, ductless core becomes a viable option from a safety point of view.	Slow blockage propagation cannot be ruled out at this time.	Establish means for detection if required.	
Sub-assembly to sub-assembly propagation (2)	Can sub-assembly meltdown propagate to its neighbours?	Establish what protection systems are required if propagation should occur.	Single sub-assembly meltdown is probably of lower probability if proper detection is used than assumed previously	Establish source terms for thermal and mechanical loads on adjacent hexcans.	Sub-assembly to sub-assembly propagation with scram may be demonstrated to be unlikely.
	With scram but with pump trip?	Enabling the consequences of the accident to be determined for risk assessment.	Sub-assembly meltdown with early scram is unlikely to lead to major core involvement.	Establish the extent of propagation if it should occur.	Sub-assembly to sub-assembly propagation without scram may lead to whole core involvement.
	At nominal power without scram?		Sub-assembly meltdown without scram may lead to major core involvement.		
	Does propagation lead to major whole core involvement?				

Extent of core damage (whole core initiators) (3)	To what degree do we proceed to whole core involvement with the commonly postulated low probability whole core initiators (e.g., LOF and TOP without scram, LOPI with scram)?	Significant core design changes may be required to limit core damage for some initiators. LOPI with scram accident may require design features to limit the depressurisation rate.	LOF with failure to scram tends to bring us to whole core involvement (true for FFTF and more so for larger systems). TOP with failure to scram may only result in limited core damage. LOPI with scram may lead to core disruption depending on design.	Time and location of cladding failure in TOP. Fuel ejection, FCI, coolant voiding, fuel sweepout, plugging, and coolability in TOP accidents. Establish margins for coolability under LOPI conditions. Establish design features required to limit core damage for specific initiators.	Current status of LOF with failure to scram unlikely to change without major design concept changes. TOP with failure to scram may only result in limited core damage, but this may not be a general result. LOPI likely to be resolved by design.
Accident energetics –Na voiding (4)	Can an overpower transient induced or enhanced by Na voiding with positive reactivity effect lead to prompt criticality with significant ramp rate? Are there any autocatalytic effects? What is the relative importance to various accident paths?	Implication on core design if positive sodium void coefficient needs to be designed out. May determine structural design basis for the plant.	Doppler effect limiting initiating accident energetics demonstrated. Na superheat demonstrated to be essentially zero for LMFBR conditions. Negative or slightly positive Na void worth, leads to benign initiating accident energetics. For cores with large positive Na void, an energetic initiating phase cannot be precluded.	Incoherence effects in voiding and clad motion both core wide and within sub-assemblies. FCI with sodium in core leading to autocatalytic effects? Fuel failure mechanisms and early fuel dispersal. Inherent mitigating effects (fission products) in the disassembly process. Probabilistic studies and target plant studies required for program guidance.	For commercial plants with large sodium void worths, the voiding induced energetics may approach or exceed containment capabilities. The uncertainties in bounding this condition may remain high. Design options which reduce the sodium void worth may provide greater margin for limited mechanical damage to primary system.

Cont'd on page 72.

TABLE 6.4 (cont'd)

Key safety issues and problems	A Brief description	B Impact on design/ licensing	C Current status	D Major remaining uncertainties	E Summary/probable resolution of issue
Accident energetics— recriticality (5)	Can high ramp rate recriticalities occur due to: (i) continuation of noncoherent initiating phenomena? (ii) re-entry of initially dispersed fuel? (iii) collapse of boiling fuel pool? (iv) autocatalytic disassembly?	If energetic recriticalities can occur, they may be difficult to contain with currently used containment concepts.	The dispersive nature of disrupted core materials is likely to preclude the occurrence of recriticality. Analyses of oxide fuel done with simplified models indicate dispersal will occur down to levels corresponding to decay heat.	Details of transition from intact to disrupted geometries. Flow regimes associated with dispersal and heat transfer in fuel–steel systems. Detail of fuel–steel blockage formation and pressurisation by steel vapour. Details of possible gravity and/or pressure driven recompaction.	Energetic recriticalities due to fuel compaction will be shown to be very unlikely.
Accident energetics –FCI–disassembly (6)	Can an energetic thermal interaction occur between fuel and coolant? With Na in? With Na out? Given the initiating conditions for a disassembly, what is the core energy release?	If energetic excursions and/or vapour explosions occur, they will determine the structural design basis for plant.	Fuel vapour expansion used as a source term energetic FCIs not considered. VENUS-II disassembly code with improved equation of state available.	Potential for energetic thermal interaction with Na with prompt burst condition. Effects of gaseous fission products on disassembly energetics. Rate and structural effects and the need for multicomponent material treatment in the disassembly.	Energetic fuel–coolant interactions will be ruled out. For high ramp rates, gaseous fission products unlikely to affect the energetics.

Issue	Questions				
System structural response (7)	What is the structural response of the reactor system to accidents? What is an appropriate structural design basis of a given reactor system?	Structural design basis of reactor must be evaluated for licensing.	Verified codes for structural response of primary vessel now available. Preliminary codes for piping system now available.	3-d codes for asymmetric systems and head response required. 3-d piping codes for analysis of loop systems required. Fracture mechanics must be considered in code application.	Ability to analyse system structural response will not pace forward.
Post-accident heat removal (8)	Are ex-vessel or in-vessel core retention systems required to accommodate postulated core meltdowns, or can this be accomplished without explicit devices for this purpose (e.g. inherent core retention, either in-vessel or ex-vessel)?	Requirements for systems to add to core retention capability will impact design. Ultimate disposition of core debris must be shown for core-disruptive accidents.	Inherent capability for PAHR in FFTF. Requirements for other systems remain to be established.	Can inherent in-vessel coolability be shown? Can inherent ex-vessel coolability be shown by passive means? Will engineering devices be required?	Combination of inherent features and /or simple design features should be adequate.
Radiological source term (9)	What radiological source term should be used for evaluating containment systems?	Level of containment/ confinement required is determined by radiological source term.	Models exist for characterising behaviour of radiological materials behaviour and transport from secondary containment to site boundary.	Models for transport and attenuation of radiological materials within primary system required. Radiological consequences associated with post-accident heat removal not well defined.	Resolution of this issue depends on resolution of energetics and PAHR issues. Engineered containment and filter systems can provide ultimate reduction in dose at site boundary to acceptable levels.

large positive sodium void coefficient and which may have commercial potential next century would be about 500 or 600 MWe. PFR, Phenix and SNR-300 provide a sensible basis for the design of a reactor of this size.

REFERENCES

1. Surrey, J., 'Some Policy Aspects of the Fast Reactor Question', this volume
2. The Council of the Pugwash Conferences on Science and World Affairs, 'Statement on Nuclear Power and Nuclear Weapons', *Pugwash Newsletter*, **14**, 11 (1976)
3. Wilson, R., *Rev. Mod. Phys.*, **49**, 893 (1977)
4. Beynon, T. D., *Rep. Prog. Phys.*, **37**, 951 (1974)
5. Long, G., *Metal Progress*, **71**, 107 (1957)
6. Cronenberg, A. W. and Benz, R., 'Vapor Explosion Phenomena with Respect to Nuclear Reactor Safety Assessment', NUREG/CR-0245 for US Nuclear Regulatory Commission, July 1978
7. ERDA 76/147, *Problems of LMFBR Steam Generating Systems*, 1976
8. Health and Safety Executive, *Some Aspects Re Safety of Nuclear Installations in Great Britain*, HMSO, London (1977)
9. Bupp, I. C. and Derian, J. C., *Technology Review*, **76**, 26 (1974)
10. Keck, O., 'Fast Breeder Reactor Development in West Germany, An Analysis of Government Policy', *Ph.D. Thesis*, University of Sussex, June 1977
11. Sir John Hill, House of Commons Select Committee on Science and Technology, Energy Resources Sub-Committee, Session 1975–76, 'Alternative Sources of Energy', *Minutes of Evidence*, 1022, HMSO, London (1976)
12. Keck, O., 'The West German–Belgian–Dutch Fast Breeder Programme: A Critical Review of Government Decision Making', *Workshop on Energy Options and Risks*, University of Chicago, Policy Centre, November 1978
13. Times Mining Correspondent, *The Times*, 30 October 1978
14. Sir Brian Flowers, 'Nuclear Power and Public Policy', Lecture re British Nuclear Energy Society, December 1976

7

The Assessment and Assumptions of Risk with Fast Reactors

Part 1: Peter J. Taylor

7.1 INTRODUCTION

Our small research group in Oxford has closely followed the nuclear power debate in Britain, central Europe, Scandinavia and the United States. It is to state the obvious to say that all matters of dispute in this debate can be reduced to a questioning of the risks and benefits, but it is perhaps not always obvious that the debate occurs on two levels and that each level has a complement of relevant facts.

The first, and most obvious, level is dispute centred upon the 'objective' facts, for example, the cost of nuclear generated electricity, the hazard potential of a certain process or the radioecological consequences of a particular discharge.

The second, less obvious, level concerns the nature of the debate itself. That is, the risks and benefits may be agreed as to fact (which is not often the case), but the evaluation differs—some groups accept the risk in relation to the benefits, others do not. Relevant facts concerning the nature of the dispute would be as follows:

(a) What groups are involved in the debate and what benefits or risks accrue to each, e.g. contracts, jobs, risk of accident or pollution?
(b) How is risk perceived and evaluated within these groups and what are the components of attitudes?
(c) What correlations are there to levels of knowledge of the technology?
(d) What procedures exist for debate, participation and information exchange and how effective have these procedures been in the past?

It has been our experience that, whenever attempts have been made to reach decisions on policy in the light of controversy on nuclear matters, the debates have been dominated by the former group of facts. Indeed, the decision-making procedure has been so structured as to preclude the relevance of the second group.

Classic examples have been the Brunner Hearings and the Windscale Public Inquiry, two procedures that set out to assess the 'objective' facts—on economics, civil liberties, pollution, etc., and to present them, with recommendations, to decision makers. The social facts relevant to the controversy were all but ignored in spite of submissions. In Brussels, Brunner, when pressed, expressed the hope that a democratic solution to the disagreement would be found. At Windscale, Mr Justice Parker declared himself at a loss to see how the detailed evidence on the social and political implications of particular nuclear options aided him at all.

In both procedures, there was no provision for the assessment of what we shall come to refer to as 'societal risk'. Consequently, by the time the 'objective' facts were re-presented (in varying degrees of completeness) to the next level in the decision-making process—either the European Commission, or the UK Parliament in these cases—important facts relevant to the social and political impact of the various options were simply not represented. Decisions were left to be made within what we would regard as a dangerously restricted framework.

It is with this general problem in mind that we now turn to the specific controversy surrounding nuclear safety, and in particular to reactor accidents that might lead to a catastrophic release—what Beyea[1] has termed 'the large consequence end of the risk spectrum'. However, we shall not be concerned to assess how credible, realistic or probable such accident sequences might be; rather, we shall be concerned to set the process of risk assessment within the broader political argument of risk and benefit. In this regard, the history of accident or safety assessment in the UK provides some illuminating examples for our general problem as outlined above.

7.2 A BACKGROUND HISTORY OF SAFETY ASSESSMENT

We are here concerned with the potential major release consequent upon a severe disruption or melting of the reactor core and subsequent loss of containment. Apart from the fire in the military plutonium pile at Windscale in 1957, when containment was at least partially effective in limiting the release, there has never been such an accident. It is at the 'least probable' or 'almost incredible' end of the accident spectrum. However, for our purposes it is instructive to note that the possibility has influenced safety and siting policy:

(a) The potential for release of radioactivity due to the thermal energy of the core was recognised at the outset (there is a large inventory of volatile radio-active nuclides), and elaborate safety precautions were taken to make such a loss of control 'incredible'. Containment and redundancy in shut-down and cooling systems were engineered such that only a series of simultaneous and highly improbable and independent faults could lead to release.
(b) However, despite the confidence of the designers and operators, these improbable events were taken into consideration in siting policy and the early Magnox reactors were all sited remote from conurbations.
(c) In the late 1960s pressure mounted to site new reactors near load centres and a review of siting policy and reactor safety took place. A Nuclear Safety Advisory Committee was formed and reported that 'the major contribution to

public safety lies in the standards achieved by design, construction, and operation of the plant', the Minister of Power then reporting to Parliament on 6 February 1968 that the new advanced gas-cooled reactor (AGR) could be operated 'nearer to built-up areas than we have so far permitted'.

At the time of this policy change, the most publicised aspect was the inherently safer design of the AGR, using prestressed concrete for the pressure vessel and internal boilers. However, from the technical literature, it is clear that it was also realised that a large release would have a similar scale of impact however remote the siting[2,3]. How far this realisation affected siting policy is not clear to us and we would invite comment.

For our present purposes, we are concerned less with the details of assessment than with the process of decision making. This process is characterised by the evolution of *expert groups* within the UK Atomic Energy Authority (AEA), Central Electricity Generating Board (CEGB) and the Nuclear Installations Inspectorate (NII) to make the assessments, the details of which are kept secret. The controlling bodies accept the assessments as adequate and the authorising bodies then act upon the assessments with regard to siting policy.

We might suppose that the public and their various representatives did not question the assessments or the benefits to be gained from taking the risks as presented to them. There was no controversy, no environmental opposition and no nuclear 'lobby' as we know it today. However, this three-stage process can be usefully summarised as it will also characterise the later debate when controversy was rife:

(1) The assessment of risk.
(2) The acceptance of the assessment.
(3) The acceptance of the risk in relation to the benefit.

It was during this period of 'unopposed' development that risk assessment evolved the technique of 'probabilistic analysis'. Farmer developed the technique as an aid to the designer in deciding what degrees of redundancy and containment it might be reasonable to incorporate[4]. He sought target probabilities of failure in relation to the consequences and in comparison to other industrial hazards.

Some doubt was expressed at the outset about designers reasoning like actuaries with regard to acceptable and unacceptable risk (see de Vathaire in the 'Discussion' in Farmer[4], p. 325), and one might be justifiably concerned that the reasoning was not fully presented to Parliament, but under the circumstances the Farmer methodology was a useful tool for which its author, in our view, is justly respected. It replaced the muddy concept of the Maximum Credible Accident and led to rigorous fault-tree analysis of accident sequences.

The Farmer curve (see *Figure 7.1*), however, has since moved out of the arena of conscientious design engineers and into the light of public debate. In so doing, it has often undergone subtle change. We need to remember, and indeed Farmer has himself sought to make this clear, that the curves are 'targets' to be achieved (see *Figure 7.2*). They are not and cannot be 'facts' of reactor operation. To understand the subtle impact such change can make, we can compare Farmer's comments on the method with more strident and less critical supporters of nuclear power.

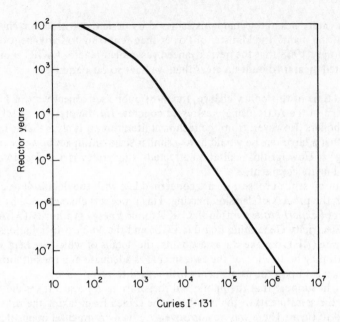

Figure 7.1 An early 'Farmer curve' (from the IAEA Symposium, Vienna, 1967): proposed release criterion.

Farmer[5], on whether the probabilities are achievable: 'an open question, there is too much credit taken for 300 reactor-years of safety' and 'a failing to see or adequately to have regard to all those minor and sometimes major features of equipment or of organisation which might nearly have led to disaster'.

Sir John Hill[6]: 'The figure (a 'Farmer' curve, see *Figure 7.3*) shows the probability of accidents of different degrees of severity from nuclear power stations compared with other accident probabilities that we live with every day.' (On the graph used by Hill, the curve was extrapolated to give 1000–10 000 casualties a probability of 1 in 10 million years, and the crucial words 'target for. . .' are omitted.)

Sir St John Elstub *et al*[7]: 'It has been calculated by Professor Rasmussen that the chances of such an accident causing fatalities for a group of 100 power plants would be one incident in 100 000 years.'

Thus, the 'target' probabilities of the Farmer method have acquired, through 'calculation', a reality, at least as far as the public face of the industry is concerned. We hasten to add that Farmer is hardly to blame for other peoples' curves.

The Flowers Report[8] presented a sound critique of probability analysis and the use of fault-trees, pointing out that not all failure modes could be foreseen nor all components ascribed a frequency of failure and concluded, 'whether the safety performance is achieved in practice will depend to some degree on factors assumed in the analysis, such as the quality of maintenance' (para. 267).

However, in contrast to this critical attitude to assessment, they were led to conclude that 'estimates (damage following an accident) *must be* presented in

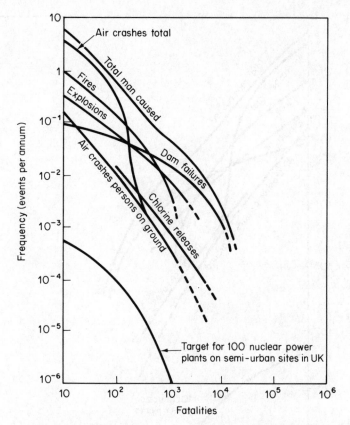

Figure 7.2 Comparison of 'target curve' with other man-made hazards (US Nuclear
Regulatory Commission (NRC), also used by the UK Royal Commission
on Nuclear Power and the Environment): incidence of man-made
disasters (in the USA).

probabilistic terms'. As we have indicated in our own studies, we believe there is
no 'must' about probabilistic presentation. Indeed, it tends to disguise some un-
pleasant facts of consequence analysis[9]. It is perfectly reasonable to present a
worst case where wind and weather combine to exacerbate the consequences of
a release (see Appendix 5). Not to present a full analysis lays one open to the
charge of deciding that which will *unnecessarily* alarm the public. At the Windscale Public Inquiry[10] (para. 11.17), Mr Justice Parker succumbed to this
temptation on the matter of releases from Highly Active Waste Storage. *Figure 7.4*
shows a detail from our report[11] on safety assessment at Windscale where weather
conditions were presented without the complexities of probability presentation[12].
If one needs to justify this approach, one need only look at the very real example
of the Amoco Cadiz disaster, where all the factors of release, weather, wind and
coastal proximity were as bad as they could be.

79

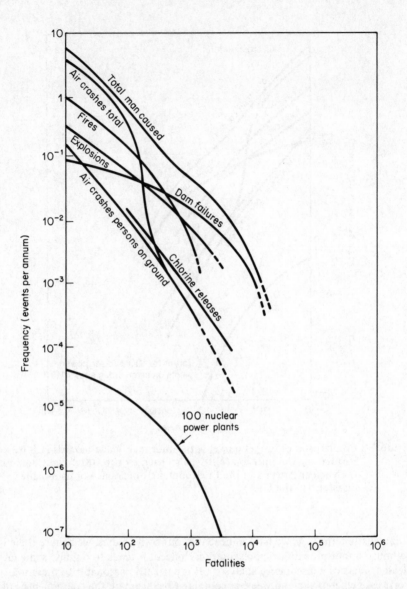

Figure 7.3 Use of the NRC curve (in *Atom*, **257**, 50) (note the omission of the words 'target for...'): frequency of fatalities due to man-caused events.

Application of this principle for the presentation of risk assessment alters the public impact of such studies quite considerably. For example, the AGR accidents considered by the Royal Commission could lead to 100 000 casualties under certain not particularly uncommon weather conditions; for the FBR study, the figure jumps to 600 000[5,12].

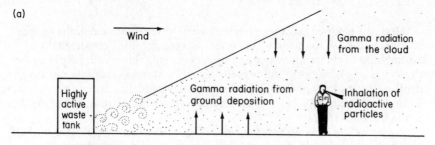

(a)

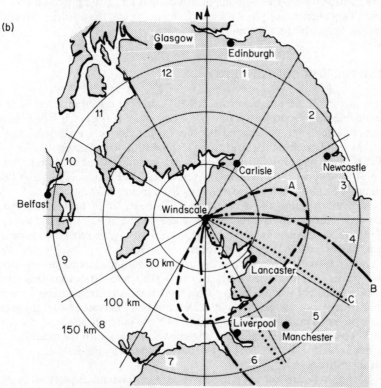

(b)

Figure 7.4 Two diagrams extracted from PERG Report OR-4 on major release
potential of Windscale. (*a*) Schematic diagram showing dispersion of
the radioactive cloud and the mechanism of irradiation of individuals.
(*b*) Boundaries of land requiring evacuation for at least 50 years due
to contamination by ^{137}Cs following release from a HAW tank.
Notes: (1) Wind is assumed to be blowing into sector 5. (2) Boundaries
A, B and C surround land where radiation levels exceed 500 mrem per
annum. (3) Boundaries B and C will extend beyond 150 km—they are
not shown here due to the limited accuracy of the computer model
beyond 150 km.

	Weather category		*Wind speed* (m s^{-1})	
A		1		1.5
B		4		6
C		6		2

81

Thus, the picture that emerges of past safety assessment, and of the present position in the UK, is one of an uncritical acceptance of the competence and assumptions of the various expert groups concerned. This is especially clear from the *Times* article[7] on risks and benefits signed by various presidents and vice presidents of the Institute of Civil Engineers, Institute of Chemical Engineers and Royal Institute of Chemistry. As we shall see, this is in marked contrast to the situation in the United States.

It may well be the case that the controlling bodies such as the NII and CEGB assessors do strongly disagree with applicants designs and assumptions—if they do so, then the arguments take place behind closed doors and the 'nuclear industrial complex' speaks with one voice on the matter.

7.3 THE PRESENT CONTROVERSY

Given the history of safety assessment in the UK *and* the operational record, it is understandable that the advocates of nuclear power are perplexed by the present debate about safety. It is understandable that they should seek explanations of motivations that have little to do with the real question of public safety. For example, Hoyle[13] and Moss[14] see the environmental objections as an orchestrated Soviet plot to undermine western energy resources. More commonly, the objectors are seen as ignorant of the real issues, having little technical knowledge, an emotional reaction against the technology, and as being led by professional agitators. We shall return to the question of knowledge and abilities, suffice it to say that the above perceptions are, in our view, wholly wrong, and that the perpetuation of these views will not aid the advocates of nuclear power.

We have detailed the structure and history of the anti-nuclear movement elsewhere[15,16]. The debate on the Continent and in the United States has seen far greater use of accident risk assessment than the UK debate, and contains a crucial stage that we may not see in Britain at all. We can best illustrate this by the events in America subsequent to the leak of WASH 740, the public hearings on the AEC and the involvement of the Union of Concerned Scientists, with the eventual production in draft of WASH 1400 for critical comment by the scientific and engineering institutions[17].

Crucial to this debate was the critical stance taken by some experts within the industry who either leaked safety reports or actually left and joined the ranks of the environmentalists. Effectively, a *re-assessment* took place, in public, of the work of the closed expert groups. Perhaps the most forthright was the report of the American Physical Society in response to WASH 1400, and the involvement of Von Hippel's team at Princeton University[18]. This involvement was able to proceed without prejudice to security or commerciality and to examine reactor safety and consequence of release in great detail.

This pattern of events has been repeated in Sweden, Denmark, Holland, Austria and West Germany, often with the funding by these states of American experts taking part in the public debate. In many cases, provision has been made for critical comment, often with detailed reports, to be incorporated in the procedures leading up to energy policy decisions. Thus, Princeton has been active in Sweden and Denmark[19], and currently in West Germany[1,20]. In the latter state, one key expert has joined the ranks of nuclear opponents; Dr Klaus Traube, chief scientist with the Euratom Breeder Programme, is now an outspoken critic[21]. Dr Traube suffered the unfortunate experience of being bugged and arrested under suspicion of terrorist links and was exonerated only by the personal apology of the Minister in charge of security before the German Parliament.

In a recent paper at the German Ministry of Technology Experts meeting on reactor safety research, Traube had this to say: 'the reference system of the nuclear power expert is restricted, in particular by limits of competence and by interiorised in most cases subconscious, preconceived understanding: the economic utilisation of nuclear power, the construction of nuclear power plants, must not be jeopardised.'

He went on to outline the history of the involvement of critical experts in West Germany, culminating in evidence to the Whyl court where the three high court judges ruled that a reactor close to Freibourg was unacceptable because of the 'residual risk'—the probability did not override the consequences, which would be a national disaster. Traube concludes that, 'The State can only do justice to its own pluralistic claims if it provides an adequate material and institutional basis for the work of critical scientists who could represent the opposition'.

This latter approach has been taken in part by the State of Niedersachsen during the licensing procedure for the Gorleben reprocessing plant. A panel of critical experts has been commissioned to provide a critical review[20]. A similar procedure was followed by the Swedes[19].

The situation in the UK is hardly comparable. There has been no major re-assessment of the work of closed expert groups—at least not in public. Critics are expected to fund their own participation and at Windscale were expected to come up with their own detailed safety analyses. Needless to say there was no access to internal reports on which to base these assessments.

The British public has hardly felt concern at the situation. It is an engaging, but to us a worrying quality of the British that they have an implicit faith in 'the scientist'[22]. We suspect this is due to the 'one voice' noted earlier, which in itself is due in part to the operation of excessive control on 'official information' and on the freedom of scientists within Government bodies to express publicly dissent from official positions. We may, therefore, never experience the stage of reassessment noted above. Much depends upon the policy of such recently formed bodies as the Commission on Energy and the Environment.

In this regard, a further stage in the foreign debate may well appear—certainly it is within the remit of the Commission. This stage, which we might dub 'justificatory' arises out of a response by the nuclear lobby to the public concern about health and accident hazards. It is characterised by quantification across the whole fuel cycle of the health and accident risks *for each energy supply option*. In the case of accident-free nuclear histories, Rasmussen-type probabilities are fed in and the results expressed in deaths, etc., per gigawatt year of power supplied. Such studies surfaced in the Ford Mitre Report and indicated that, for example, coal stations were more 'hazardous' than reactors[23]. Although it should be pointed out that one of the assumptions was—no reprocessing.

A more recent and thorough investigation has been presented by the Atomic Energy Board of Canada[24]. This study concludes that nuclear is safer than all supply options except natural gas, including solar space heating. The chief components of risk from non-conventional sources are materials acquisition, fabrication and construction which involve more man hours of hazardous employment, as well as necessary inputs from conventional back-up systems.

The report is not beyond criticism. Again reactor accident frequencies are assumed which may or may not be borne out in practice, but certainly it is an important step towards the quantification and comparison of risks. An extension of this procedure is currently in progress in the United States—the Committee on Nuclear and Alternative Energy Systems[25]. [Their comments on breeder safety are referenced here as they concluded that containment should be engineered to withstand the worst core-disruptive accident even though they regarded this as highly improbable.]

If such studies are to be instigated in the UK for the purposes of energy planning, then quite considerable changes in reactor safety assessment will have to be made. AGR and breeder fault analyses will have to be publicly presented and discussed before figures can be accepted universally for further consideration. If critical experts are to be expected to produce safety analyses, as at Windscale, then they will have to have at least access to official information[26].

Certainly a start has been made in this direction by Mr Benn at the Department of Energy. The NII/NRPB consequence analysis for a notional breeder accident is a detailed and forthright document, but what of a detailed safety analysis? As we noted at the outset, acceptance of the risk analysis will be a phase of any assessment in public. The RCEP noted this crucial point: 'the consequences of such an accident (reactivity with gross fuel vaporisation) would be so severe that the FBR could scarcely be contemplated on an extensive commercial basis unless it is established with a high degree of confidence that (these accidents) can be controlled' (para 304).

Our emphasis would therefore be on the 'establishment of confidence' and we would hold that this is scarcely possible within the present framework of risk assessment by closed expert groups.

Already there are significant doubts about past UK reactor safety assessment following events at Hunterston and Hinkley Point. At Hinkley Point, the newly commissioned AGR lost both primary and secondary cooling following a pipe break and simultaneous failure of a seawater valve in the pump house[27]. Cave and Holmes placed the probability of this event at 10^{-5} per reactor life[2]. They also stated (p. 277) that the AGR was a good deal safer than the PWR in this loss-of-coolant situation, 'due to the heat capacity of the moderator and boilers there is a period of 3–4 hours available in which to restore core flow and boiler feed flow and thus to prevent the melting of any cans of fuel'.

At a recent Oxford meeting, Dr Franklin of NPC was visibly embarrassed when asked the calculated probability of the Hinkley incident. He told us that in any case there had been standby fittings for fire hoses which were connected up almost immediately. This is at variance with the advice to the Secretary of State, who in a reply in Parliament on 2 February 1978 stated that a period of 3 hours elapsed before cooling was restored—the connections for fire hoses were not present on the Hinkley plant. It may well be that the Minister's statement that there was no danger of a release would have applied to a fourth hour had there been one, but we raise the most serious doubt that this can be assured without a detailed and public presentation of the safety report.

Only thus can we obviate a danger recognised by Professor Schäfer at the recent Bonn Reactor Safety meeting[28] : 'that the risk analysis may be employed as some kind of "numerical magic" for the hypnotisation of objectors, plaintiffs and, in particular, judges and public decision makers'.

Schäfer also stresses the use of comparative studies and of the need to understand the limitations of such comparison. It is to these limitations of the methodology we have so far discussed that we now turn.

7.4 SOCIETAL RISK

In our introduction we stated that those involved in the nuclear debate tended to ignore certain very real factors concerning the nature of the debate itself. We stated that such factors as the attitudes and perceptions of some of the protagonist groups would have a bearing on the outcome of the debate. We can use the issue of risk assessment to illustrate this point, and raise a series of questions:

(a) How is quantification viewed by the public, especially where different kinds of deaths, diseases, or risks are given similar import?
(b) What components of risk carry greatest weight in the formation of attitudes?
(c) How are attitudes distributed within concerned groups and how are these affected by information programmes?

It is interesting to note that these questions were first raised by researchers commissioned by the IAEA from the International Institute of Applied Systems Analysis (IIASA) to look into social and political aspects of the nuclear debate. The problems of risk assessment *and acceptance* (i.e. divergent evaluations) are thus comparatively well researched but little publicised. Pahner, for example, concluded that the public perception of nuclear risk was quite unique, embodying as it did, 'a symbolic threat of death' linking a galaxy of subconscious fears of nuclear explosion, cancer and genetic damage[29]. The Batelle Institute in Frankfurt is currently reviewing this field of risk perception and assessment for the German Government[30]. We have presented our detailed views elsewhere and concluded that public 'lay' perceptions are likely to be quite different from those of schooled technologists or legal judges[15]. In our view, a meaningful public participation will have to come to terms with this problem, for it may well be the case that a large proportion of the populace, perhaps a majority nationally, certainly majorities around some projected installations, are immune to certain arguments and will not find nuclear power acceptable.

That the controversy surrounding our example of safety assessment may be irresolvable is indicated by recent IIASA work from Stallen and Meertens, two Dutch social psychologists[31]. They suspect that there exist within the population two quite different evaluatory frameworks for all technology. There is a polarity, in their view, between the 'technocratic' and the 'interactive' orientation, the former viewing each new technological development as progressive if it increases control over the environment, and often, almost any increase in complexity and sophistication may be, of itself, progress. The latter orientation evaluates technological change according to the degree of freedom it allows for human interaction—

hence it is strongly influenced by values particularly relating to community.

We would submit that Stallen and Meertens are correct, but further, that the 'interactive' view is widespread and particularly present in the social movements stemming from the late 1960s that now manifest themselves in the 'ecological movement', the 'peace movement' and the growing calls for decentralisation and participatory democracy. This we have documented[15]*.

The strength of this movement is hard to assess. Certainly not all those concerned about nuclear power are part of a social movement, but as we have documented, the movement has considerable resources and is able to present persuasive argument in public forums, courts and licensing procedures. In the case of the UK, it will be able to highlight scientific controversy and possibly affect the British public's view of safeguards and experts. In such areas as Galloway, the Cheviots and Cornwall, intervenors and critical experts will find ready political acceptance. The recent referenda in Austria and Belgium and the problems of Scandinavian Governments show clearly that the arguments have considerable force.

However many votes are counted one way or another, we would urge that one factor be not forgotten—the strength of feeling and commitment of the nuclear opponents—for, as we have seen in France and Germany, this will determine the extent of the response needed to press ahead with opposed developments. The Malville Breeder site, for example, was protected by troops using explosive grenades and one protestor was killed by them. At Kalkar, the Government had to mobilise thousands of heavily armed police and armoured columns. There is a social barrier to the FBR, the surmounting of which may require the permanent institution of such methods, and before we are tempted to dismiss such opposition as the work of agitators, as did the recent MP for Berwick at Torness, we might consider that some at least are guided by the motivation expressed by Professor Rotblat, that 'we must abolish nuclear energy or it will abolish us'.

We thus present for consideration the concept of a 'societal risk'. It can be expressed as a question: what political means will have to be developed to overcome the social barriers to nuclear development?

7.5 RECOMMENDATIONS

We have elsewhere offered constructive proposals towards an overhaul of the decision-making procedures concerning nuclear and general energy policy[32]. These were formulated with the FBR Inquiry in mind. We regard the institution of public debate along the lines we have outlined in this paper as an essential prerequisite for open and democratic decision making. The controversy may not be resolvable, in that opposing groups will always dissent, but it is right that everyone concerned should have a full appreciation of the risk of all options.

Specific to the FBR, a number of inadequacies need to be remedied before informed debate on risks can proceed:

*For a review of this material, see Political Ecology Research Group forthcoming publication PERG RR-4.

(1) A thorough safety study must be presented by the applicants for consideration, in draft, by institutions and interested bodies. A critical review should be funded and published by Government.

(2) A risk assessment should take place for a series of energy supply options, to include quantification of the health effects, in particular from pollution, but including an analysis such as that of the Canadian AEB.

(3) A programme of research into public attitudes and perceptions of risk, and in particular on the social and political impact of each energy policy decision, is long overdue.

Such a programme will take several years and cost a great deal, but we would submit, not as long as breeder R & D and perhaps 1% of the cost.

REFERENCES

1. Beyea, J., 'Reactor Safety Research at the large consequence end of the risk spectrum,' *BMFT Experts Meeting on Reactor Safety Research*, Bonn, September 1978
2. Cave, *et al.*, 'Suitability of the AGR for Urban siting,' *Proc. Symp. IAEA*, SM 89/32, Vienna 1967
3. Gronow, W. S., 'Application of Safety & Siting Policy to Nuclear Plants in the UK,' *Proc. Symp. IAEA*, SM 117/21, Vienna 1969
4. Farmer, F. R., 'Siting Criteria—a new approach,' *Proc. Symp. IAEA*, SM 89/34, Vienna 1967
5. Farmer, F. R., 'The development of adequate risk standards,' *Proc. Symp. IAEA*, Jülich, Vienna 1973
6. Hill, J., Lecture to the Royal Society for the Encouragement of Arts and Manufactures in Commerce, *Atom*, March 1978
7. Elstub, S. J., *et al.*, Article in *The Times*, 6 February 1978
8. Royal Commission on Environmental Pollution, 6th Report, *Nuclear Power and the Environment*, HMSO, London (1976)
9. Political Ecology Research Group, *A Study of the consequences to the public of a severe accident at a commercial FBR located at Kalkar, West Germany*, PERG RR-1, January 1978
10. *The Windscale Inquiry*, Report by Mr Justice Parker, HMSO, London (1978)
11. Political Ecology Research Group, *The Windscale Inquiry and Safety Assessment*, PERG OR-4, August 1978
12. Kelly, *et al*, *An Estimate of the radiological consequences of notional accidental releases of radioactivity from an FBR*, NRPB R-53, August 1977
13. Hoyle, F., *Energy or Extinction*, Heinemann, London (1977)
14. Moss, R., Article in *Daily Telegraph*, 25 September 1978
15. Taylor, P. J., Evidence to *The Windscale Inquiry*, Day 96, 1977
16. Political Ecology Research Group, *Nuclear Power in Central Europe*, PERG OR-1, 1977
17. Reactor Safety Study, *An Assessment of risks in US Commercial Nuclear Power Plants*, WASH 1400, USNRC, October 1975
18. American Physical Society, Study Group on LWR Safety, *Rev. Mod. Phys.*, 47, Suppl. No. 1 (1975)
19. Beyea, J., *A study of the consequences of hypothetical reactor accidents at Barsebeck*, Report to the Swedish Energy Commission 1978

20. Hirsch, H., 'Public Participation and Nuclear Energy,' *The Gorleben International Review*, Paper to UNESCO Conf., Vienna, November 1978
21. Traube, K., 'Man Machine Interaction', *BMFT Experts Meeting*, 1978
22. White, 'New Society Special Survey on Nuclear Power,' *New Society*, **39** No. 756, March 1977
23. Ford Foundation, *Nuclear Power Issues and Choices* (1977)
24. Inhaber, H., *Risk of Energy Production*, AECB-119/Rev, 1 May 1978
25. Avery, *et al*, *Report of the LMFBR Safety Subgroup of the Risk/Impact Panel of CONAES*, June 1977
26. Thompson, G., Evidence to *The Windscale Inquiry*, Day 96, 1977
27. Eadie, A., 'Parliamentary Reply,' *Hansard*, 2 February 1978
28. Schäfer, R., 'Risk Analysis,' *BMFT Experts Meeting*, 1978
29. Pahner, P. D., *A psychological perspective of the nuclear power controversy*, IIASA RM-76-67 (1976)
30. Conrad, J., *Zum Stand der Risikoforschung, Kritische Analyse der theoretische Ansätze im Bereich des Risk Assessment*, Batelle, Frankfurt (1978)
31. Stallen, *et al, Draft Report* to IIASA (in press)
32. Political Ecology Research Group, *Public Participation and Energy Policy*, PERG OR-6, May 1978

An Alternative View—Some Safety Considerations of the Fast Breeder Reactor

Part 2: F. R. Farmer

7.6 WHAT IS SAFETY?

It is generally recognised that absolute safety cannot be achieved and that most industrial activities carry some risk of causing harm to people and the environment.

The aim of design and operation is to keep the risk sufficiently low, and there will be argument as to what this implies. The increased recognition of risk has rendered obsolete the concept of maximum credible accident and currently leads more to a discussion of residual uncertainties in accident and consequence modelling rather than a prolonged description of the devices installed to minimise the chance of accidents and a demonstration of their reliability. In following this trend, i.e. to discuss uncertainties, it should be understood that in many ways uncertainties will always remain, it will only be possible to establish an increasing degree of confidence that the system and its behaviour is well understood. This is important when considering the early versions of new systems—reactors or aircraft; it is not possible to establish the same degree of confidence in the first as may be developed by the fifth or the tenth, and to accelerate this learning process there should be extensive exchange of safety information and collaborative programmes on safety issues on an international scale. To date, collaboration has been good; it is to be hoped that commercial considerations will not prevent this exchange.

7.7 BRIEF COMPARISON OF FAST AND THERMAL REACTORS

The fast reactor core is much more compact than thermal reactors, and hence the heat production per unit length or volume is greater. The cooling system has to cope with this; sodium is a very effective heat transfer medium operating at near atmospheric pressure.

The control of power in a fast reactor is easier than most thermal reactors. Both depend on the small fraction of delayed neutrons. The more compact core of the fast reactor simplifies the procedures for controlling power because the effectiveness of any control rod extends virtually throughout the reactor and is only slightly affected by the position of other rods, so that elaborate systems of sector control are not needed as in large thermal reactors.

It is easy to provide redundancy in the control and shut-off systems; a large reactor may have 30 rods, the insertion of any five would shut down the reactor. The aim in current designs is to achieve sufficient diversity.

The removal of heat from the fuel is in some ways easier in a fast reactor and in other ways more difficult than water or gas-cooled reactors, as indicated by the following examples.

As sodium reacts violently with water—and additionally, hydrogen as a neutron moderator must be kept away from the core—all designs have intermediate sodium-to-sodium heat exchangers and the final sodium-to-water exchangers are placed some distance from the reactor. On the other hand, the normal operating temperature of the sodium is several hundred degrees below its boiling point, so that there is a wide margin to accommodate temporary disturbances of power or flow. In reactors of the tank design, the whole primary circuit is immersed in several thousand tons of sodium; as a result, the fission product decay heat can be absorbed for several hours, sodium flow being by natural circulation, i.e. without pumps. It is yet to be shown that the transition from the flow conditions under power to convective flow of a shut-down reactor can take place safely—this is still being investigated.

7.8 FALSE FEARS OR INCORRECT IMPLICATION OF FAST REACTOR DANGERS

In brief, it has been said that a fast reactor may behave like an atom bomb; that fast neutrons make the reactor more difficult to control; that the reactor is not in its most reactive configuration. The fuel of current fast reactors contains plutonium plus a higher percentage of ^{238}U—this is a neutron absorber. This fuel cannot explode like a bomb.

89

The microsecond lifetime of neutrons in a fast reactor, as compared with milli-seconds in a thermal reactor, has no bearing on normal operation, in which control operates through the delayed neutron fraction, and under most accident conditions the response is not dominated by the neutron lifetime, but by the Doppler coefficient of the U/Pu fuel which cancels most changes in reactivity—which may result from some accident condition—by a corresponding rapid change in the fuel temperature. The reactor becomes less reactive as the fuel temperature increases.

It has frequently been said that the particular feature of a fast reactor is that the core is not in its most reactive configuration. This is true, but is an over-simplification in comparison with thermal reactors. The light-water reactors in their normal operation are not in their most reactive state—but for different reasons. The water reactors can be rendered prompt critical by changes in their coolant; the PWR by a rapid temperature drop of the water; the BWR by collapse of the steam voidage in the reactor core. Both conditions are now deemed to be reduced to an improbable event or one of low probability by design measures. The same degree of confidence has not yet been established for the fast reactor, perhaps because designs are viewed more critically. There are two types of fault conditions peculiar to a fast reactor which increase reactivity, one by the rapid expulsion of sodium from the mid-core regions, or secondly by a collapse or compression of the core into a slightly smaller volume

7.9 A POSSIBLE SAFETY PRESENTATION

As said earlier, it is not, and never will be, possible to show that accidents cannot happen—it is only possible to show that the accidents are unlikely and their consequences reasonably understood and tolerable in relation to their estimated frequency.

How might this be done? It is likely that in newly developed systems such as the fast reactor, two lines will be followed in parallel:

(1) To show that effective measures are provided to prevent the escalation of minor faults into major accidents. Hence all major accidents have a very low probability of developing.
(2) To show that even if the preventative measures fail, the ways in which accidents develop, and are limited and contained, are sufficiently well understood through an adequate, even if not perfect, appreciation of the phenomena involved.

How far can these lines be followed at present? The normal classification of accidents into errors of control, heat removal, etc., appear to be well covered with redundant and diverse protective equipment. The adequacy of these provisions may be subject to some discussion, mainly to improve confidence in the lower levels of accident frequency. It seems that two troublesome areas remain—a degree of uncertainty in the structural integrity of the internal reactor structures, when requiring confidence at low frequency levels, and secondly some uncertainty in the various possible modes of failure propagation within fuel sub-assemblies and their subsequent development.

Currently, many methods of early identification of onset of failure are possible, but the application of all sensors could lead to over-instrumentation to an extent that the reactor could never operate effectively.

These uncertainties are likely to be resolved—not solved—with sufficient confidence to enable the next stage to proceed. The greater stumbling block—ideologically—lies in the second parallel line of agreement—that if preventative measures fail, the accident can still be contained. The difficulty in this approach is the continued escalation of the assumptions defining the accident; there is seldom an upper limit; there is no maximum credible accident as conveniently postulated for the development of the light-water reactors. Rather than question too closely the initiating assumptions, it may be more instructive to study two important phenomena:

(1) the mode of fuel behaviour and core dispersion in a 'whole-core accident', and
(2) the conditions of violent or passive interaction between hot fuel and sodium—known as FCI (fuel–coolant interactions).

The 'whole-core accident' has been under study in the USA, the UK, France, Germany, Japan and the USSR for up to 20 years. The research and theoretical programmes have been, and continue to be, extremely large and complex. The object has been to find out how fuel behaves when heated rapidly in the range 2000 to over $4000\,^{\circ}$C in time scales from a few milliseconds to a second or so.

Various core-dispersive mechanisms exist. Fundamentally, the core must disperse if the fuel is vaporised.

This mechanism can be violent and is studied to find the dynamic and static loadings in the core region and its surroundings.

For some accidents, molten fuel may be swept out of the core by the coolant; the loss of a small fraction of fuel would terminate the reactivity excursion.

A third dispersive mechanism exists through internal gas pressures generated in the fuel by fission products. Theoretically, this could be very effective and could prevent fuel vaporisation, or reduce its extent. This mechanism has only been studied in the last few years, but experiments are continuing.

Fuel–coolant interactions have also been studied for well over 10 years, using various materials, e.g. tin, aluminium, uranium oxide, water, sodium, freon, etc. Some theoretical correlations have been suggested in recent years, but there is not yet a generally agreed model and more work is required particularly with UO_2 and sodium. The question arises, when will there be sufficient understanding of 'whole-core accidents' and 'FCI', and what constitutes 'sufficient understanding'?

Let us return to the proposition of a two-pronged approach to safety:

(1) To show that accidents can be identified at an early stage and safely terminated. This is likely to be developed with considerable confidence.
(2) To study accident conditions assuming protective measures fail—this route is open-ended; at no time will it be shown that any accident developing in any way with no safeguards can be tolerated by any one design.

However, these studies should lead to some optimisation in core materials, core geometry and surrounding containment, so as to tolerate a wide range of possible accidents even if these are of low probability.

This present stage in safety analysis is described by Hafele[1] as a third stage in the safety debate. It is instructive to follow his argument.

The first phase from 1944-1959 gave much attention to the short neutron life-time, the relatively small fraction of delayed neutrons, the speed of control devices, a positive power coefficient, etc. This arose because the reactors were small and used metallic fuel.

The second phase up to 1970 considered larger reactors with mixed oxide fuel. The debate was then around the sign and size of the Doppler coefficient, then around superheat, the sodium void coefficient and FCI.

The overriding feature of the third phase is a widespread and multilevel approach of proving the elements of the design concept that arose from the second phase. In this connection, one has to realise that the nature of what is considered a proof has evolved very substantially in the years since 1970. Engineering judgement was no longer considered sufficient. The probability approach to reactor safety was introduced and related to the concept of residual risks, extremely small in probability but very large in the perceived consequences.

The focus of the scientific and technical attention of the third phase is therefore the depth and credibility of the experimental and theoretical proof of the design concept. In many countries, there is now a growing interest in the attempt to quantify confidence and the degree of uncertainty. This is likely to continue.

REFERENCE

1. Hafele, W., *et al*, 'Fusion and Fast Breeder Reactors', *IIASA Report* RR-7-8, International Institute for Applied Systems Analysis, A-2361 Laxenburg, Austria, November 1976 (revised July 1977)

8

The UK Fast Breeder Programme

Walter Marshall

In this contribution to the debate on the fast breeder programme, I shall confine myself to a factual and simple statement of what fast breeder reactors are, how they work and how their operation is intimately linked to the operation of the fast reactor fuel cycle. I shall keep this presentation as simple as possible. I do not

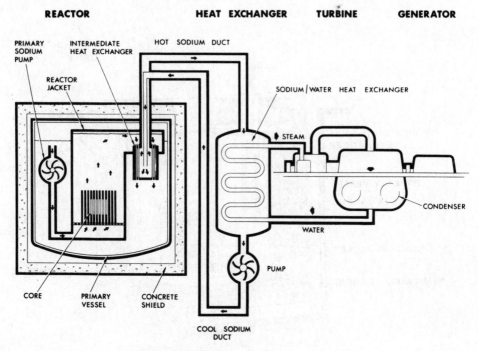

Figure 8.1 Outline of a fast reactor.

apologise for that; there is so much about fast breeder reactors which, in general discussions, is not well understood or is misrepresented that it seems worthwhile establishing some simple and relatively straightforward points as soon as possible.

Figure 8.1 shows a simplified outline of a fast reactor. The fuel elements, containing a mixture of plutonium and uranium oxide, are placed in the core of the reactor where the fission process develops heat which is carried away from the core by molten sodium. That hot sodium is pumped through into heat exchangers so that its heat is exchanged with the sodium in the secondary circuit and this, in turn, exchanges heat with water in the secondary heat exchangers which, therefore, produce steam to drive the turbines and produce electricity. Those of you who are familiar with the operation of thermal reactors will notice that the fast reactor has one additional circuit which is intermediary between the primary sodium circuit and the final water/steam circuit. The purpose of this intermediary sodium circuit is easily described. In the primary circuit, the sodium becomes slightly radioactive and, therefore, we have judged it not prudent to use that primary sodium directly in a heat exchanger with water but, instead, interpose an intermediate sodium circuit. This additional circuit and the use of sodium carries the immediate implication that the capital cost of a fast reactor will be somewhat larger than that of a thermal reactor.

However, this additional capital cost must eventually be compensated for by a saving in fuel costs. This saving in fuel costs comes about because, once launched, a fast reactor needs no significant input of new fuel because it has the remarkable

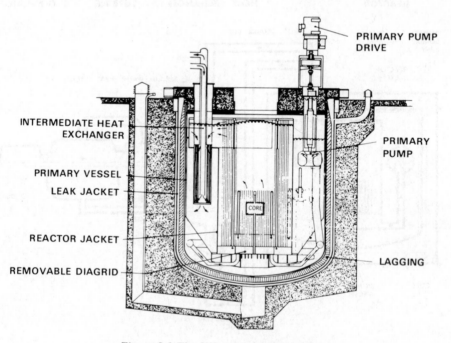

Figure 8.2 The UK sodium-cooled reactor.

94

ability of breeding its own fuel from what would otherwise be a waste material. For technical reasons, the fast reactor can also achieve a high 'burn-up' of fuel.

To understand this last point, we need to look more carefully at the core region of a fast reactor which is shown in outline in *Figure 8.2*. That figure shows that the true core of the reactor has fuel elements made of mixed plutonium/ uranium oxide clad in stainless steel, and that the true core is surrounded by a region we call the blanket which is filled with blanket elements made from depleted uranium oxide clad in stainless steel.

The plutonium needed to launch the fast reactor and the depleted uranium needed in the blanket, both come from the thermal reactor programme. Thermal reactors produce plutonium in the course of their operation and they also produce uranium depleted in the uranium-235 isotope. This depleted uranium appears as a reject waste material either directly from our Magnox stations or as a reject material from enrichment plants.

During the operation of the fast reactor, some of the plutonium in the true core of the reactor is incinerated and some is both produced and incinerated *in situ*. The neutrons which escape from the core are absorbed in the blanket, and, in that absorption process, uranium-238 is converted to plutonium-239. By designing the fast reactor carefully, it is possible to arrange that the production of plutonium in the blanket exceeds the incineration of plutonium in the core and the reactor can then be called a 'breeder' because, once started, it is able to make its own fuel and perpetuate itself, i.e. 'breed' itself indefinitely provided it is fed with enough depleted uranium-238—which is a plentiful but otherwise useless waste product.

It is obvious from this brief description that the UK fast breeder programme can be divided into two parts: that which is concerned with the operation of the reactor itself and that which is concerned with the fuel cycle to launch and maintain it. I shall now discuss these two items in turn.

Our experience on the operation of fast reactors themselves comes from our early work with the Dounreay fast reactor and the more recent work with the prototype fast reactor (PFR). The PFR commenced operation in 1974 and, since that time, the nuclear reactor itself has given us no significant problems. There have been no fuel failures and no problems in handling the sodium, the primary and secondary circuits have both performed excellently, the mechanical sodium pump, the intermediate heat exchangers, rotating shields, fuel-handling equipment and a mass of newly developed instruments were all brought together for the first time in the PFR and have worked well. The sodium pumps have operated continuously for four years with an availability close to 100%; the fuel-handling machinery was used to load the first core within a month or so of putting sodium into the reactor and that machinery immediately achieved the design specification; the operation of the nuclear reactor itself over the whole power range has been completely straightforward. The early months of operation of the PFR were plagued with small leaks between the sodium and water sides of the secondary heat exchangers. Those leaks have been traced back to difficulties in the initial manufacture of the heat exchangers and we have now devised techniques to detect defective welds in advance. This experience with the secondary heat exchangers, though frustrating at the time, has in retrospect been a very valuable experience because we are now confident that we know how sodium-water heat exchangers should be designed, manufactured and tested. The PFR has also had a large number of teething troubles connected with the conventional electricity generating plant. These troubles have been neither worse nor better than those experienced on non-nuclear power stations.

Given this experience, which is similar to that being obtained in parallel in France and in Russia, we can make the judgement that there are no fundamental engineering problems connected with the construction or operation of the reactor. The next tranche of information to come from the PFR will concern the perform-ance of the fuel. The fuel elements put into the reactor initially have not yet achieved their designed burn-up but should do so sometime during the next two years. We should then be able to satisfy ourselves about the burn-up actually achievable in practice, but I expect the fuel performance to be entirely satisfactory and, therefore, I shall say nothing further either about that or about the engineer-ing of the fast reactor. I shall, therefore, turn to another area of fundamental importance, namely the operation of the fast reactor fuel cycle which is getting increasing attention within the UK fast breeder programme.

I mentioned earlier that the reactor was called a 'breeder' because it could be operated to produce more fuel than it consumed. It is also called a 'fast' reactor

TABLE 8.1 THERMAL REACTOR (PWR)

Plutonium (kg per GW yr)			
Creation	710		
Destruction	380	Input	0
	330 =	Output	330
		PWR production	330
		Magnox production	600
		Candu production	600

TABLE 8.2 FAST REACTOR

Incineration in core (kg per GW yr)		*Production in blanket* (kg per GW yr)	
Input	2800	Input	0
Creation	530	Creation	479
	3330		479
Destruction	750	Destruction	70
Output	2580	Output	409
Incineration =	Input − Output	Production =	409
=	220		
Balance = 189 kg per GW yr			
Net incineration without blanket is 220 kg per GW yr		Maximum production with blanket is 189 kg per GW yr	

96

because the neutrons are not slowed down (i.e. thermalised), as they are in a thermal reactor. Because these two descriptions of the reactor are important, it can sensibly be described as a 'breeder fast reactor', and this would have been a wise nomenclature because fast reactors are not necessarily breeders. However, it has become conventional to use the adjectives in the opposite order and, thus, the reactor is described as a 'fast breeder'. Unfortunately, this phrase has been interpreted, in the non-technical literature, as applying to a reactor which breeds plutonium copiously and rapidly in the way that rabbits are fast breeders of rabbits. In this sense, the fast reactor is, in fact, a very poor breeder of plutonium. It can produce it quite well in the blanket but it incinerates so much in the core during the course of operation that the balance between incineration and production is somewhat delicate. Because these points are so important and so widely misunderstood, I think it is valuable to give a balance sheet for plutonium production and incineration in the various types of reactors and this is set out in *Tables 8.1* and *8.2*.

Table 8.1 shows the balance for a PWR. This table shows that a 1000 MW PWR produces 330 kg of plutonium per gigawatt year (i.e. operating for a full 365 days). In comparison to this, an equivalent size of Magnox or Candu station would produce about 600 kg. All these reactors, therefore, which we call 'converters', use uranium-235 and convert it to electricity and produce quite large quantities of plutonium. The equivalent table for the fast reactor is more complex because I need to show the figures separately for the core and for the blanket and these are shown in *Table 8.2*. The precise numbers in these tables are dependent on the detailed design of PWRs and fast reactors, but the numbers given will serve as examples. From all this, we can deduce several simple characteristics of fast reactors and they are as follows:

Fast breeder reactors:
(1) do not breed fast,
(2) use fast neutrons and breed slowly,
(3) mostly incinerate plutonium,
(4) have a core which is an incinerator,
(5) have a blanket (optional) which is a producer, and
(6) the balance between incineration and production is delicate.

Despite these points, it is usually represented in newspaper articles and simplified accounts that the fast reactor in some way or another produces plutonium copiously—the reverse in fact true. It incinerates plutonium efficiently and its operation can be optimised to produce *more* plutonium, to hold the total amount of plutonium in *balance* or, indeed, simply to *incinerate* it.

In *all* circumstances, for a given production of electricity from nuclear power, the use of fast reactors instead of thermal reactors must *decrease* the amount of plutonium in the world.

These simple tables have also made it clear that the fuel cycle of the fast reactor is much more intimately linked to its operation than the thermal fuel cycle is linked to the operation of the thermal reactor. The reason for this is quite simple. After the fuel has been used in a thermal reactor, there is no immediate use for it, whereas, after the fuel and blanket have been used in a fast

97

reactor, there is an immediate need for that fuel and blanket material to be reprocessed so that the plutonium they contain can be used to maintain the reactor in operation.

This comment serves to introduce the fact that there are several fuel cycles which involve uranium and plutonium. Each have their own characteristics and it is very important that these should be clearly understood and distinguished one from the other. I shall distinguish four fuel cycles and these are described in *Figure 8.3* in an outline form.

The first fuel cycle shown in *Figure 8.3*, the once through fuel cycle, is that which most countries in the world are now operating. The uranium fuel is put through a thermal reactor and is then stored, together with its plutonium content, without reprocessing. This is done either because reprocessing capacity is not available or, in the case of the United States and Canada, because it is judged to be premature and unnecessary. The situation here in the UK is different from the rest of the world because our thermal reactors are gas-cooled and the spent fuel from a gas-cooled reactor is not easily stored for long periods of time. In our case, therefore, the fuel must be reprocessed for environmental reasons to separate out the plutonium, the uranium and the fission products.

The second fuel cycle is that where the fuel is reprocessed, the plutonium is extracted and is refabricated into fuel for the thermal reactor. This is called thermal recycle of plutonium. It is the fuel cycle which a large number of countries are planning to operate in the near future. The exceptions are the United Kingdom and France, which believe it to be both unnecessary and an unwise use of plutonium, and the United States which considers it to be a proliferation risk. The economic attraction of recycling plutonium into thermal reactors is very simply stated. It replaces the need for some fresh uranium fuel and could be applied at a relatively early date because thermal reactors are already available.

The third fuel cycle is that to *launch* the fast reactor and uses the plutonium made in thermal reactors. In this fuel cycle, the spent fuel from a thermal reactor is reprocessed and the plutonium is stored until it is needed. Notice that, in these three fuel cycles, there is no urgency to get the plutonium fabricated into fresh reactor fuel because in the once through fuel cycle, the plutonium is not used at all, in the case of thermal recycle of plutonium, the value is marginal unless the cost of uranium increases substantially and, in the third case, fast reactors are simply not yet available to use the plutonium.

The fourth fuel cycle is that used to maintain fast reactors in operation and it is the development and proving of that fuel cycle which is now absorbing a large part of the UK fast breeder programme. That fuel cycle contains three main parts.

(1) The reprocessing of the core and breeder elements to extract the fission products from them and recover a mixture of uranium and plutonium oxides in the correct proportions to make up a fresh charge of fuel for the fast reactor core.
(2) The fabrication of fresh fast reactor fuel elements from this recycled material.
(3) The disposal of the fission products (and actinides) as high-level waste.

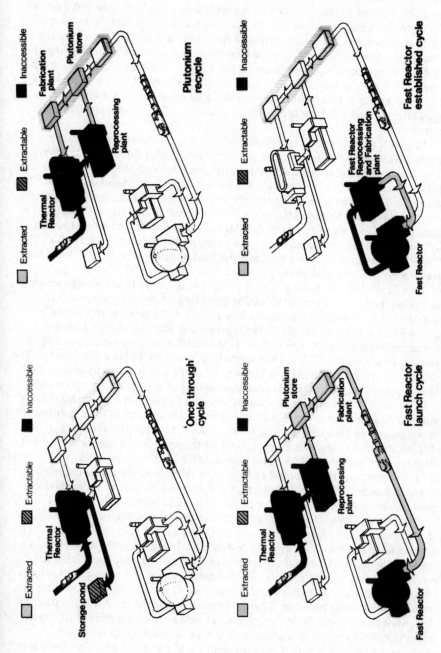

Figure 8.3 Outline of four distinguishable fuel cycles.

Extracted · Extractable · Inaccessible

Thermal Reactor

Storage pond

'Once through' cycle

Extracted · Extractable · Inaccessible

Thermal Reactor

Reprocessing plant

Fabrication plant

Plutonium store

Plutonium recycle

Extracted · Extractable · Inaccessible

Thermal Reactor

Reprocessing plant

Plutonium store

Fabrication plant

Fast Reactor

Fast Reactor launch cycle

Extracted · Extractable · Inaccessible

Fast Reactor Reprocessing and Fabrication plant

Fast Reactor

Fast Reactor established cycle

The waste disposal process is very similar to that used for thermal reactors, so I shall say nothing about it. However, it is important to recognise that the reprocessing of the core and breeder fuel and the refabrication into fresh fuel is very intimately linked to the operation of the system as a whole because it is essential, for good economic operation, for the plutonium to be extracted from the spent fuel and returned to the reactor within a period of, say, two years, to keep the total plutonium inventory at a reasonable level. Typically, therefore, fast reactor fuel will be reprocessed, at the latest, about 12 months after it comes out of the reactor—which is as Magnox fuel is reprocessed nowadays. This is in contrast to the reprocessing of fuel from a water-cooled thermal reactor system where, typically, the fuel will have been allowed to cool for five to ten years before reprocessing.

In addition to these differences of emphasis, the fast reactor processing plant itself is significantly different from the reprocessing plant for thermal reactors. The percentage of plutonium in the material going through the plant is much higher than it is in thermal reactors, but the total bulk of material is lower in quantity. These plants are designed with such a geometry that a criticality accident is impossible. Therefore, the overall size of a fast reactor reprocessing plant is smaller, the pipes are of narrower diameter and the various components have a different geometry compared to a reprocessing plant for thermal reactor fuel. These differences are not of a fundamental kind but they need to be defined and the process needs to be demonstrated and fully proven.

Here in the UK, the spent fuel that comes out of the PFR will be reprocessed in a plant shortly to be commissioned at Dounreay. Many years ago, we decided that the most economical and common-sense way to do this was to dismantle, decommission and decontaminate the reprocessing plant we already had at Dounreay for a different type of fuel. That plant has now been rebuilt to be suitable for fuel from the PFR and will come into operation within a few months. The decision to set about this plant in this way has been well justified by events. De-Decommissioning, decontamination and rebuilding of the reprocessing plant cost only a small fraction of what it would have cost to have built a new plant. We are also planning to build a fuel fabrication plant at Dounreay so as to demonstrate, in one place, the complete fast reactor fuel cycle. However, the exact design of that plant and the choice between various routes towards fuel fabrication cannot be made until we have had more experience of fuel performance in the PFR.

It is appropriate that I should end this contribution by discussing the relationship which some people see between fast reactors and nuclear weapons proliferation. However, I shall be brief because Dr Jasani is addressing this subject. First, we should note that different countries put quite different emphases upon the use of plutonium as a nuclear fuel. Here in the UK, when we refer to plutonium fuel we invariably associate that with the fast reactor. However, in the USA there is a strong tendency to think of it in terms of the recycle of plutonium to thermal reactors and they usually relegate the commercialisation of fast reactors to the remote future. Inevitably, therefore, a good deal of the debate on this subject has tended to be at cross purposes. There is a tendency for the US Administration to make statements about the use of plutonium which they have derived from consideration of the recycle of plutonium into thermal reactors. There is then a tendency for us to interpret those statements in terms of the use of plutonium in fast reactors and, in that context, we find the statements difficult to comprehend. This confusion between the use of plutonium in thermal and fast reactors has been made more complicated by the domestic USA position where President Carter is opposed to the Clinch River fast reactor project and it was not apparent, initially, whether his opposition to that was due to a hostility to fast reactors in general or to that project in particular.

Because of this, in all discussion about proliferation questions, it is therefore prudent to be meticulous in specifying which fuel cycle is being discussed. The fuel cycle to launch fast reactors has a limited life span and, therefore, the most important of the fuel cycles to consider is, in my opinion, that needed to maintain the fast reactor in operation.

It is my judgement that if there are proliferation dangers in that fuel cycle they can be met by a combination of technical and institutional means. That appears to be the position of the US Administration also. To quote Mr Joseph Nye[1] : 'Consequently, we carefully avoided criticism of other countries' breeder reactor research and development programs. We believed that there was time to develop information, technology and institutions for a safer fuel cycle before breeders became commercially competitive. On the other hand, recycling plutonium for use in light water reactors represented a clear and present danger with at best marginal economic advantages. By challenging assumptions about recycling while keeping an open position on the breeder reactors, the Administration sought to buy time and to focus on an area where a consensus might be developed.'

I will also draw your attention to the very careful and precise speech made by Mr Joseph Nye in his talk to the Uranium Institute where he again made a clear distinction between the use of fast reactors and the recycle of plutonium to thermal reactors. A careful study of that speech satisfies me that the differences between our viewpoints are primarily matters of emphasis and timing.

It seems, therefore, that the British and American points of view are not as different as they are sometimes represented to be by the more strident commentators in the press. Since America speaks with many voices, it is worth pointing out that Mr Nye stated in his talk that he was speaking officially for the US Administration, and was not just expressing his private view. It is my belief that any remaining differences will be resolved in the INFCE discussions.

REFERENCE

1. Nye, J., 'Non-Proliferation—A long term strategy,' *Foreign Affairs*, April 1978

9

Radiation Hazards: Areas of Uncertainty

Patricia J. Lindop

In the area of radiation hazards it is particularly important to know, as nearly as possible, what we do not know.

In the context of today's discussion, we need to look at the possible effects of our uncertainties about radiation hazards, not as they apply to the practice of radiological protection in the nuclear industry today but how they apply to the policies and plans for a nuclear power programme in the next century. These uncertainties will require answers whether we are in favour of or against the development of the fast breeder programme. If we are unsure of our assessment of radiation risks, we could err on the side of caution to the extreme of making the development of an 'acceptably safe' industry either technologically or economically impossible. However, the same uncertainty about radiation hazards could, in times of economic difficulty, make vulnerable standards of radiological safety, and lead to the piecemeal erosion of necessarily stringent controls.

I here limit my questions to the biological effects of radiation on man, with emphasis on occupational exposures, and to the late effects of radiation on the individual and the population. These include both somatic and genetic damage.

Somatic effects are of two main types—first, degenerative damage (so-called non-stochastic effects), such as fibrosis in the lung or dermatitis in the skin, and, secondly, carcinogenic damage (stochastic effects) which is seen as an increased incidence of all types of cancers, the seriousness of each depending on which type of cancer occurs in which organ.

I am not going to discuss acceptability of risks or cost–benefit analysis, as these are value judgements; the whole of society must help to make these, not an individual expert in any one field. What, then, are the bases for assessment and control of radiation hazards?

The first is an understanding of *dose*, that is, the amount of energy deposited in the tissues by ionising radiation, and the factors which change the biological effectiveness of the dose, such as dose rate and the quality of the radiation. We need to know whether the radiation is distributed sparsely over a long pathway in the tissues as with X- or gamma-rays, or whether it is concentrated within a short track length and is therefore more likely to deposit a significant amount of energy in a biological target, resulting in damage. The latter is called *high LET radiation* of which the alpha-particle is an example. This may be 10 to 20 times more effective in producing damage than gamma- or beta-rays.

The second basis is the *response* to be measured. This will depend on the tissues irradiated, and within those tissues the cell populations at risk for either cell killing or cell mutation with its potential for producing cancer. Thus, the basis depends on our understanding of the relationship between the dose and the response and *Figure 9.1* shows some of the forms which this can take. The simplest is the linear relationship, where the observed effect increases in proportion to dose.

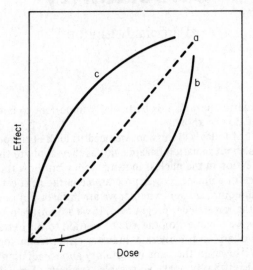

Figure 9.1 Three possible relationships between dose and effect.

The curve 'a' is of the form established for genetic damage in laboratory animals and is the one used for assessment of cancer risks in man. It implies that any increment in dose is harmful and that there is no threshold or safe dose for that effect. In contrast, curve 'b' shows no detectable effect until an apparent threshold dose has been exceeded and gives the impression of a safe dose, *T*, below which no harm is done. Curve 'c' shows a steep initial slope at low doses which is decreasing at higher doses, implying some interaction at these higher doses of the cancer induction effect (which would be expected to increase with increasing dose), and some other, as yet not understood effect, e.g. cell killing, which would decrease the number of cells capable of expressing the cancer.

Our first uncertainty is which of these curves best fit the data for man and, secondly, how will the fit be modified by the quality of radiation, gamma-rays compared with alpha-particles and by dose rate.

The main sources of human data are shown in *Table 9.1*. Each population gives us some data on different modes of exposure, whether as a single whole-body dose

104

TABLE 9.1 POPULATION DATA USED FOR RISK ESTIMATES

Nuclear weapons	A-bomb survivors (Hiroshima and Nagasaki)
Nuclear tests	Marshallese Islanders Other fall-out
Medical radiotherapy	Spondylitics Other non-malignant
Medical diagnosis	Fluoroscopy (TB) Thorotrast Pelvimetry (foetus)
Occupational hazards	Uranium and other miners Dial painters Radiologists
Natural environment	Areas of high background

of external irradiation, of predominantly gamma-rays or neutrons, as in Hiroshima and Nagasaki, or internal irradiation, as with the Thorotrast patients or uranium miners. The National Radiological Protection Board, which is responsible in the UK for making recommendations of basic standards for radiation protection, feels that from several of these epidemiological studies the frequencies with which cancers are induced at high doses are known within a factor of 2, and that for low doses the rate of induction per rad will be between one-half and one-quarter of these for most radiations. For alpha-particles and neutron irradiations, the rates per rad at low levels are likely to equal those at high levels. It is on these data that the International Commission on Radiological Protection thinks that now we have got a fairly realistic estimate of the cancer hazard.

But how sure are we of these estimates? We take some examples in areas which might be of relevance as far as occupational exposure is concerned. *Figure 9.2* shows an analysis of the data from people who had incorporated predominantly radium-226—a bone-seeking isotope—into the body. It shows an increased incidence of cancer of the bone after an estimated dose of irradiation accumulated throughout life. It gives the impression of a threshold dose. It is seen, however, that above this threshold dose there is no significant increase in cancer incidence with increase in dose over two orders of magnitude! It was on the basis of these data that the maximum permissible body burden for a bone-seeking isotope, radium-226, was set, and this is our basic reference level. A more recent study shows that in patients treated with radium-224, which is again a bone-seeking isotope but has a shorter half-life and therefore delivers the dose at a faster rate, it will also produce cancer of the bone but with a linear dose response. From these data, *Table 9.2* shows that there is a difference in sensitivity between children compared with adults, but more markedly that, contrary to our theories of the dose rate independence of alpha-particle irradiation, there seems to be an inverse dose rate effect as far as cancer induction is concerned, i.e. spreading the dose out over a longer time is *more hazardous*.

Now if radium-226 is our basis, then we need to know how other bone-seeking

105

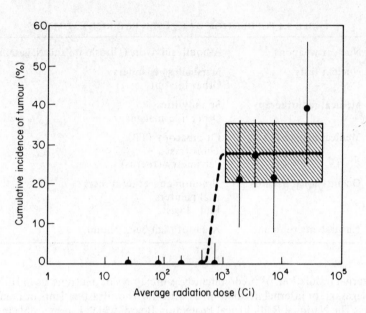

Figure 9.2 Incorporation of radium-226 into the body[3].

TABLE 9.2 BONE TUMOURS—INDUCTION PER 10^6 PER RAD ALPHA[5]

Radium-224	Children	40 (short)
		220 (long)
	Adults	30 (short)
		170 (long)
	Low LET to endosteal cells 1–5	

isotopes are likely to compare in toxicity with that for radium. One of these with which we are most concerned, because of the large volumes in which it may be handled is plutonium. *Table 9.3* gives a list of different people's estimates of the relative toxicity of plutonium-239 compared with radium-226. It is seen that these estimates differ by several fold.

A further problem is that bone-seeking isotopes do not necessarily just produce cancer of the bone. For example, *Table 9.4* shows a summary of the data that we have from patients who had been injected with Thorotrast, which is a radioactive contrast medium used in diagnosis. Approximately 4000 of these were followed up for possible incidence of bone cancer. And here we see a major lesson that we still must learn, and that is, in this field of hazard assessment, we must look for effects other than the ones we expect. For instance, here if we were looking at the data only for bone cancer there are few, but if we look at the incidence of marrow-related diseases both of cancer (leukaemia and myeloma) and of the failure of

TABLE 9.3 ESTIMATES[7] OF RISK FACTOR FOR BONE
CANCER—PLUTONIUM-239 RELATIVE TO RADIUM-226

Source	Risk factor
ICRP	5 [2]
MRC (Spiers)	8 [8]
Marshall and Lloyd (Dog RBE-6)	18 [9]
Marshall and Lloyd (Dog RBE-16)	48 [10]
Morgan (Dog RBE-16)	64 [11]

TABLE 9.4 'SKELETAL' TUMOURS—THOROTRAST
(after Mole[8])

n	~ 3900
Bone sarcoma	(?) 6
Leukaemia	50
Multiple myeloma	5
Myeloid dyscrasia	26

bone marrow, we can see that there is a much higher incidence of these and therefore a significant additional hazard to that of bone cancer.

Thorotrast is a bone-seeking isotope, which is distributed also in the bone
marrow and other organs, and gives an analogy with the possible effects of the
distribution of plutonium in the body. *Table 9.4* shows that with Thorotrast
there is a *leukaemia* hazard.

Some of the best data we have from the populations in *Table 9.1* are for the
incidence of leukaemia, because leukaemia is one of the earliest forms of cancer
to occur after irradiation. From a study of the survivors of Hiroshima and
Nagasaki, we can learn some facts about the radiation dependence for leukaemia
induction. *Figure 9.3* shows the incidence of leukaemia in survivors from Nagasaki.
In Nagasaki the exposure was to a single external irradiation, predominantly
gamma radiation, that is low LET, and there is apparently a decreased effectiveness of the irradiation with decreasing dose, i.e. a curvilinear threshold type of
dose response relationship. In contrast, *Figure 9.4* shows that the incidence of
leukaemia following irradiation at Hiroshima, where there was a significant
neutron, that is a high LET, component, there is no threshold, no safe dose, and
the induction per rad for high LET irradiation is similar for high and low doses.
If we summarise the risk rates that one can derive from looking at Hiroshima and
Nagasaki, and looking at patients with irradiation of the bone marrow who have
been treated for ankylosing spondylitis, Thorotrast populations, and children who
have been treated for tinea capitis with irradiation of the scalp, there is some sort
of concordance for the risk rate of leukaemia from these populations as shown in
Table 9.5. From several different sources there is apparently some reliable estimate
of risk rate for leukaemia, which is, of course, a very rare form of cancer.

However, if we look at another form of cancer, cancer of the lung, the

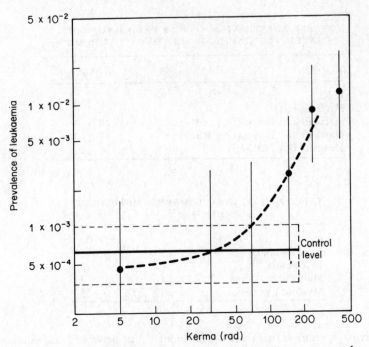

Figure 9.3 Incidence of leukaemia in survivors from Nagasaki[4].

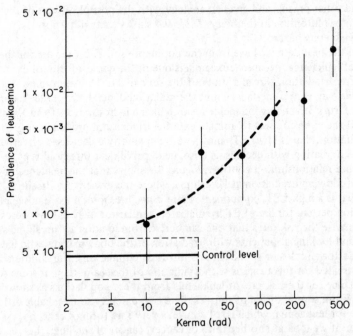

Figure 9.4 Incidence of leukaemia following irradiation at Hiroshima[4].

TABLE 9.5 LEUKAEMIA–INDUCTION PER
10^6 PER RAD[5]

Hiroshima and Nagasaki	30
Radiotherapy	11–25
Thorotrast alpha	50–55
Children (tinea)	30
Low dose	Low LET 15–25

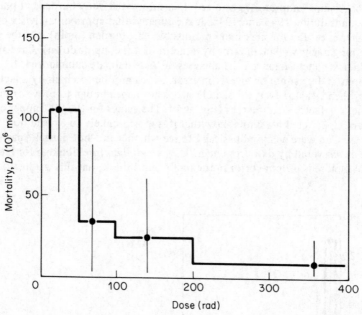

Figure 9.5 Lung cancer–dependence on dose[5].

Hiroshima data do *not* show a decreasing effectiveness of the irradiation as the dose is decreased. *Figure 9.5* shows that the maximum efficiency, that is per unit of irradiation, is in the low dose range with a decreasing effectiveness in the high dose range. Now if we have this change in effectiveness of irradiation per unit dose at high doses compared with low doses (*Figure 9.1*, curve 'c') linear extrapolation from high dose data will, of course, *under*estimate the risk at low doses. The lung cancer risk rates for Hiroshima and Nagasaki are in fact much lower than has been found for uranium and other miners, as shown in *Table 9.6*. This might well be because of the greater effectiveness of internally incorporated radioactivity in uranium and other miners, where the irradiation can be localised to the bronchial epithelium but *we do not know*.

Moreover, it could be argued that the higher risk rate for the miners is due to the non-uniform dose distribution–a problem we turn to in the 'hot particle' discussion.

109

TABLE 9.6 LUNG CANCER–INDUCTION
PER 10^6 PER RAD[5]

| Hiroshima and Nagasaki | 10–25 |
| Uranium and other miners | 40–180 |

There has been a great deal of discussion as to which irradiation data we should use for estimating risk rates for cancer of the lung, particularly for plutonium. One useful set of data currently available is that relating to a group of 25 workers from the Los Alamos Laboratories who acquired substantial body burdens of plutonium compounds during 1944 and 1945. It is estimated that approximately six of these may have exceeded the maximum permissible lung burden (mplb), but the dose to the lung might well be in error by a factor of 4 (i.e. up or down)! Cave and Freedman's conclusions[1] on the analyses of these data are summarised: '. . . 0.016 μCi mplb ^{239}Pu cannot be greatly in error. . . . cannot be too high by a factor of greater than 14. However, the sample size is not large enough to prove or disprove whether the factor is, in fact, as large as 14. The values for the 99% upper confidence and upper credible limits show that it is most unlikely to exceed 60.'

These data were analysed basically to see whether the 'hot particle' problem could be answered by data from man. They were, therefore, looking for factors of tens of hundreds or more error in the mplb, and I think that this is a justified use.

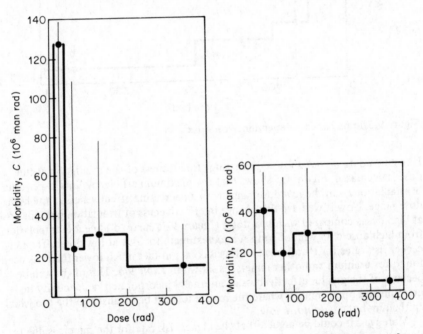

Figure 9.6 Cancer of the breast–effectiveness of the irradiation on dose[5].

However, these data have also been repeatedly implied to justify the current maximum permissible lung and body burden for plutonium. We cannot in a situation, with this size of error, say that the maximum permissible lung burden is adequately low!

Other data I would like to discuss in support of our doubts are those for cancer of the breast because the breast is probably one of the most radiosensitive tissues as far as cancer induction is concerned. *Figure 9.6* shows that for Hiroshima and Nagasaki we have again got this picture of a relatively high efficiency per rad for induction at low doses with a falling off in this efficiency at high doses. Therefore, *Table 9.7* shows that use of the Hiroshima and Nagasaki data might well underestimate the cancer risk rate as far as breast is concerned.

TABLE 9.7 WEIGHTING FACTORS FOR
INDIVIDUAL ORGANS[9]

Tissues	Weighting factor
Gonads	0.25
Breast	0.15
Red bone marrow	0.12
Lung	0.12
Thyroid	0.03
Bone	0.03
Remainder	0.30
Total	1.00

One explanation for this low risk rate from the Japanese data is that the natural incidence of cancer of the breast is lower in Japan. This raises another area of uncertainty. How much does the risk rate depend on the natural incidence of the cancer? Does it mean that we will need to have different standards in different parts of the world depending on what is the natural incidence of the disease?

So far we can see that there are several forms of cancer on which we have got data, and it is on the basis of these data that ICRP has put into its new recommendations a weighting factor for the risk rate, or sensitivity to cancer induction in different organs as shown in *Table 9.7*. The 'implied' dose equivalent limits for individual organs may be found by dividing the annual whole body limit (50 mSv per year) by these weighting factors. This seems a sensible adjustment to make from a practical point of view where several different routes of exposure might be involved. It does, however, lead to the surprisingly large increases in the new dose limits to different organs, as shown in *Table 9.8*, particularly for organs such as bone, bone marrow and lung, where we have got very little confidence that we can increase the dose of radiation to these without increasing the damage. Here, too, is introduced a dose limit for bone surfaces which is for a non-stochastic effect limit. (In brackets is the limit based on cancer risk.) I have been unable to find the data for which non-stochastic effect on bone the ICRP have based this figure.

But our real concern in today's discussion is not only with the best estimates

TABLE 9.8 DOSE EQUIVALENT LIMITS[10]

Tissue/organ irradiated singly	mSv per annum	
	ICRP 26	ICRP 9
Whole body	50	50
Gonads	200	50
Breast	330	150
Red bone marrow	417	50
Lung	417	150
Bone surfaces	500*	Bone 300
	(1670)*	

of risks for low LET irradiation, but for high LET irradiation, particularly when incorporated in the body. This is because radioactivity in huge quantities is associated with all parts of the fuel cycle.

Plutonium is an alpha emitter about which we are concerned because it is to be processed in relatively large quantities. But not only is it plutonium but other actinides such as americium and curium which will be of particular concern. How hazardous are these?

Table 9.9 shows that the estimated excess annual mortality from a maximum permissible body burden of 0.04 μCi plutonium-239 is nearly an order of magnitude greater than for 0.1 μCi radium-226.

TABLE 9.9 ANNUAL EXCESS MORTALITY ESTIMATE[7]
DUE TO 45 YEAR EXPOSURE TO BODY BURDENS OF
Ra-226 AND Pu-239.

Organ	Excess mortality per 10^6 persons	
	0.1 μCi Ra-226	0.04 μCi Pu-239
Bone	74	427
Marrow (leukaemia)	26	70
Liver	–	350
Remainder	2	6
	~ 100	~ 850

Table 9.10 shows the comparison of skeletal burdens for plutonium-239 compared with americium-241 and curium-244 and there is at least a two-fold difference in toxicity. Therefore, we should be concerned about having large quantities of americium and curium around.

If we are to use the plutonium already in store from the thermal nuclear programme, one option, or perhaps necessity in the first cycle, will be the extraction of americium. This of course, without adequate containment, could be a source of occupational and population hazard. Can we adequately contain these highly radiotoxic elements?

TABLE 9.10 COMPARISON OF ANNUAL EXCESS CANCER MORTALITY ESTIMATES FOR CONTINUAL INTAKE FROM AGE 20 TO 65 SO THAT THE ICRP MAXIMUM SKELETAL BURDEN IS REACHED AFTER 50 YEARS[7]

	Annual mortality per 10^6 persons		
	^{239}Pu skeletal burden 0.036 μCi	^{241}Am skeletal burden 0.0355 μCi	^{244}Cm skeletal burden 0.0315 μCi
Bone	179	189	240
Marrow (leukaemia)	34	35	45
Liver	171	360	700
Remainder	7	2	10
Total	~ 400	~ 600	~ 1000

TABLE 9.11 ESTIMATES[5] OF COLLECTIVE DOSE COMMITMENT

Step in fuel cycle	Collective dose commitment (man rad per MWe per annum)
Mining, milling and fuel fabrication	
(a) Occupational exposure	0.2–0.3
Reactor operation	
(a) Occupational exposure	1.0
(b) Local and regional population exposure	0.2–0.4
Reprocessing	
(a) Occupational exposure	1.2
(b0 Local and regional population exposure	0.3–0.6
(c) Global population exposure	1.1–3.3
Research and development	
(a) Occupational exposure	1.4
Whole industry	5.2–8.2

If they are contained within the nuclear plant, is this likely to increase the hazard to those occupationally exposed? If they were released into the environment, is it going to cause any further trouble?

UNSCEAR (1977)[5] estimated that the collective dose commitment from the nuclear energy industry is about 7 man rad per megawatt (electric) year (*Table 9.11*).

With the present installed nuclear energy capacity of 108 GWe and a world population of 4 billion, the radiation dose which each of us receives, due to nuclear energy production is 0.2 mrad per year. This is only one quarter of a per cent of the dose from the natural background, and clearly is quite negligible.

113

But then the contribution which nuclear energy makes at the present time to the world's energy consumption is also negligible, as it is less than one per cent. If, as is projected, nuclear power becomes the major source of energy in the world (since oil will be exhausted, and it will become less acceptable to mine coal, and—as is claimed—alternative sources will never make a significant contribution), the production of nuclear energy will have to go up by several orders of magnitude. If it should go up only by a factor of 200, as IAEA a few years ago predicted would occur by 2020, the additional radiation dose could become half of the natural background. It is unlikely that this would be acceptable.

Such an expansion of nuclear energy could come only through fast breeders, since with thermal reactors the uranium resources would have long been exhausted. With slow breeders using fast neutrons there would be a negligible dose commitment from uranium mining and milling. There could also be reduction in the dose commitment from research and development. On the other hand, the UNSCEAR table did not include the dose commitment from waste disposal, nor did it allow for minor accidental releases, of the type encountered at Windscale, and which are bound to occur with the widespread use of nuclear energy. The UNSCEAR estimate of the dose commitment is therefore likely to be correct.

We have therefore to conclude that, quite apart from other problems to be encountered from the fast breeder (ecological, proliferation), the radiation hazard to the population may prove to be the factor limiting the utilisation of nuclear energy.

But, in the area of uncertainty, we even now need to consider occupational exposures in the fuel cycle with some concern. *Table 9.11* shows that occupational doses are the major proportion of the total dose commitment. Estimates by the American Physical Society put the occupational proportion as four times larger than the general population dose. They point out also that, since the dose rate is higher in occupational exposure, the relative biological effectiveness is likely to be higher. The number of workers in the nuclear industry is a tiny fraction of the world population. It follows that this group carries a disproportionately large burden, many orders of magnitude higher per capita than the population at large. The occupational dose limit is 50 mSv (5 rem) per year with an average dose claimed to be one-tenth of this. From the UK experience, *Figure 9.7* shows that in a reprocessing plant a significant proportion of those occupationally exposed are receiving between 3.5 and 5 rem external irradiation per year, and together with this one must put the internal irradiation which is not included.

And it is as far as the reprocessing plant is concerned that we are going to have the largest potential exposure in terms of the slow breeder fast neutron programme. There was already in the early 1970s an anticipated need to increase the decontamination factors if the through-put at Windscale were to be increased, and as shown in *Table 9.12* it is seen that as far as alpha-particles are concerned the normal limit is about 6000 Ci per year and the achievable containment factor in the 1970s was 90. This needed to be increased, if the discharge limits were to be contained, to about 8000 in the year 2000. The problem that arises with the introduction of the slow breeder fast neutron programme is shown in *Table 9.13* where, in fact, because of the need to reprocess, there will be a through-put of plutonium, americium and curium in quantities per GWe as large as about 5×10^5 Ci of plutonium and approximately 4×10^4 Ci americium, and this is in fuel

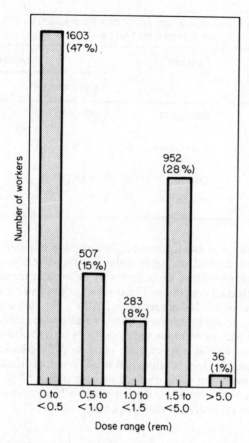

Figure 9.7 Doses received by fuel reprocessing workers in 1975[6].

TABLE 9.12 ESTIMATES[11] OF LIQUID EFFLUENT DECONTAMINATION FACTORS
TO BE ACHIEVED TO MEET PRESENT DISCHARGE LIMITS AT WINDSCALE

Nuclide	Present limit (Ci per annum)	1970 (1)	1980	1990	2000
Sr-90	30 000	230 (1100)	800	3 000	4 000
Ru-106	60 000	300 (660)	1 100	13 000	35 000
Total β	300 000	500 (1200)	1 600	12 000	30 000
Total α	6 000	40 (90)	1 200	6 000	8 000

reprocessing alone. Together with that, one has got to take into account the
amount of plutonium, which is 1757 kg per year, to be handled in fuel fabrication.

If we are to keep in balance Dr Walter Marshall's proliferation security arrangements in terms of burning up plutonium in a fast reactor programme, we must
consider, in the context of radiation health, the actual volume of radioactivity

TABLE 9.13 ANNUAL AMOUNTS OF TRANSURANICS IN FUEL CYCLE
OF A 1000 MWe FAST BREEDER[12]

Occurrence	Element	Amount	
		(kg per annum)	(Ci per annum)
Fuel reprocessing	Plutonium	1970	$4.49 \times 10^5\ \alpha$ $1.30 \times 10^7\ \beta$
	Americium	17.8	$3.82 \times 10^4\ \alpha$ $1.87 \times 10^3\ \beta$
	Curium	0.676	$1.12 \times 10^6\ \alpha$
Fuel fabrication	Plutonium	1757	

which we are likely to be moving around. This position of concern is even more reinforced by the estimated volume of alpha-particle waste arising from an initial fast reactor programme. These alpha wastes are in terms of megacuries per annum. I know that the claimed containment factors for radioactivity are now seen to be needed at the level of 2×10^9 according to IIASA[2], but I am concerned as to whether these can or will be achieved. If not, we are talking about radioactive substances which become incorporated in the body, distributed we do not know how, with an unknown but long retention in the body, and biological effectiveness for harm approximately 20 times that of the low LET irradiation for which estimates of risk have been made. Perhaps we have time to think again whether we should go into a programme where there is potentially a long-term commitment to a hazard which we are uncertain that we can control.

REFERENCES

1. Cave, L. and Freedman, L., *Proceedings of Symposium on Transuranium Nuclides in the Environment*, IAEA, 1976
2. Hafele, *et al*, *Fusion and Fast Breeder Reactors*, IIASA, Vienna, 1978
3. Evans, R. D., Keane, A. T. and Shanahan, M. M., *Radiobiology of Plutonium*, Salt Lake City, J. W. Press (1972)
4. Rossi, H., Personal Communication, 1974
5. United Nations Scientific Committee on the Effects of Atomic Radiation, Report, *Sources and Effects of Ionizing Radiation*, 1977
6. Rotblat, J., *Bulletin of the Atomic Scientists*, 1978
7. Ellett, W. H., Nelson, N. S. and Mills, W. A., *Proceedings of Symposium on Transuranium Nuclides in the Environment*, IAEA, 1976
8. Mole, R. H., *Health Physics,* July 1978
9. International Commission on Radiological Protection, *ICRP Publication 26*, Pergamon Press (1977)
10. National Radiological Protection Board, *Publication,* NRPB R63, 1977
11. Martin, A. and ApSimon, H., *Proceedings of Symposium on Management of Radioactive Wastes from Fuel Reprocessing*, OECD, 1973
12. *Health Implications of Nuclear Power Production*, WHO Regional Publication, European Series No. 3, 1978

Part III
Economic Appraisals

10

The Economics of Coal and Nuclear Power Plants

Michael J. Prior*

10.1 INTRODUCTION

This paper is largely based upon a comparison of electrical generating costs from coal-fired power plants and thermal nuclear reactors carried out in our group during 1978. Reports on this work will be published in 1979 and will give full details as to methodology and source data. The general scope of this area of Economic Assessment Service work is to establish the competitive position of coal-based energy in various end-use applications in the industrial, domestic, commercial and transport sectors. Electricity is one important form of coal-based energy; others include gas, liquid fuels of various types, and direct coal combustion. These are all being evaluated in various EAS studies.

The primary task of this stage of work was to establish the competitive position of coal using current technology, which, in the case of electricity production, means pulverised-fuel (PF) combustion for solid coal and light-water reactor (LWR) systems for nuclear power. Various advanced coal systems are being developed, notably fluidised bed combustion (FBC) and combined-cycle gasification, and an EAS study is under way to evaluate these. Some results of this study are presented here, to illustrate the kind of cost advantages which may be obtained, but no detailed comparisons are made.

In the case of nuclear power, two alternative thermal reactor systems have been used for commercial power generation, the British advanced gas-cooled reactor (AGR), and the Canadian Candu reactor. Very little useful economic data are available for these reactors. The AGR has been plagued with design and construction difficulties and no long-term operating experience has accrued, whilst the fuel costs of the Candu are based upon a unique natural uranium/heavy-water combination and are difficult to evaluate.

The main conclusion of British studies is that AGR electricity costs are likely

*The work for this paper was done while the author was with the Economic Assessment Service, IEA Coal Research. He is now a senior consultant with ERL Energy Resources Ltd.

119

to be rather higher than LWR costs[3]. It may be assumed that results derived for LWR power plants are unlikely to be improved by AGR units.

The final section of this paper deals with the economics of the fast breeder reactor (FBR), in particular the liquid-metal FBR (LMFBR). This work is an extension of previous EAS work and should be regarded as much more tentative than the LWR figures. No good cost data are available for the FBR, so that any economic analysis must be rather insecure. The intention of this section is to set up general economic performance criteria which *ought* to be met by any commercial FBR, to assess the scanty FBR cost data against these criteria, and to raise, but not answer, certain problems about the systematic analysis of any proposed FBR development.

10.2 METHODOLOGY

The basic calculating procedure used was to find the levelised electricity unit revenue required over plant lifetime to pay off all capital and operating costs and interest charges. The computational tool used was a flexible discounted cash-flow project assessment programme developed at the Oak Ridge National Laboratory.

The assessment basis was that of real-resource analysis, that is, no account was taken of taxes or of any effect induced by purely financial factors such as debt/equity ratios, and so on. This corresponds to an assessment carried out taking the national economy as the basic unit and fits in well with the assessment schemes usually used by national planning bodies.

The initial calculations were performed on constant capital and fuel costs for both coal and nuclear systems, though the effect of various cost changes was included both as a parameter and in sensitivity analysis. It is, of course, unlikely that either coal or nuclear plants will not be subject to cost changes over their lifetime. However, it is, in our view, necessary to establish certain base-line costs before subjecting the analysis to the effect of such changes. This is particularly relevant to the effects of fuel cost changes where many analyses fail to distinguish adequately between the effects of general and of specific cost inflation. The former make no difference to the results of investment appraisal made at this level, that is, the national economy. The latter, of course, may make a substantial difference.

The study was made using two internal discount rates, 5% and 10%. One of the general weaknesses of all studies of this kind is that there is no *a priori* reason for preferring any particular discount rate. This problem becomes particularly acute when comparing projects, such as these, in which there is a clear-cut capital/operating cost trade-off. In these cases, decisions of fundamental importance can appear to be reversed on the basis of apparently arbitrary changes in discount rate. The use of a low discount rate, perhaps only 3–4%, has been argued by those who view energy supply as an essentially social function with the requirement to use long time horizons. On the other hand, the use of low discount rates on the supply side, in distinction to relatively high rates on the demand side, leads, in principle and possibly in practice, to misallocation of resources.

This is not the place to make an extended discussion of the discount rate problem. However, three points are significant:

120

(i) Although the UK Treasury discount rate is currently 5%, and this is supposed to be the test rate for all energy projects in the nationalised sector, this will leave a very large number of energy investment decisions which are made using higher discount rates. This includes both the bulk of North Sea oil and gas investment and most industrial and private consumer 'point-of-use' applications. It is an open question how much misallocation of investment results from this difference, but some must occur.

(ii) A constant discount rate makes it difficult to allow for risk. Many observers might feel that there is a significant difference in risk between, say, the Hereford industrial CHP scheme and the Torness AGR. The former is based on well proven reliable technology, the latter on a less proven process. Yet both are assessed using a common discount rate with no apparent allowance for risk.

(iii) A constant discount rate should be applied consistently. This means that, for example, when the Treasury discount rate dropped from 10% to 5%, the marginal cost of coal from new mines would drop, in line with the cheaper cost of capital. In fact, of course, this did not happen, as the NCB prices coal in accordance with financial targets and market pressures, not Treasury discount rates.

The cost analyses described in this paper all rely upon published data though, whenever possible, these have all been reduced to a common basis by adjusting factors such as contingency and architect/engineer fees. The estimates have all been adjusted to a common price basis by the use of a standard cost index[1]. In cases where major cost discrepancies have remained after this, the cost concerned has normally been used as a parameter in subsequent analysis. Much of these data were based upon non-British sources. There is, however, no reason to believe that major discrepancies exist. As will be seen, coal plant capital costs tend to be consistent internationally, whilst suggested UK LWR costs are close to the average of international values.

10.3   COST BASIS

10.3.1   Capital Costs

The capital cost of LWR generating plants is an uncertain and disputed figure. In the course of preparing this work, planning estimates from reputable bodies were obtained as high as £630/kW and as low as £210/kW. (Both these sources released data only on an unattributable basis.) *Table 10.1* shows figures recently published by various European and US sources. British estimates are not immune from this variability; the National Nuclear Corporation report of 1977[2] suggests a range of £291–342/kW for a PWR, the Energy Commission paper[3] on coal/nuclear costs uses £407/kW. This difference has never been adequately accounted for.

The wide uncertainty which exists over nuclear capital costs can be attributed to a number of factors, including;

(i) failure of commercial bodies to charge a 'full price' for initial reactor orders;
(ii) varying degrees of optimism over the out-turn of nuclear construction costs;

121

(iii) differing safety requirements between countries; and
(iv) the use of different allowances for cost inflation during a period of rapidly escalating prices.

It is difficult to quantify any of these, though opinion and rumour are easily obtainable. One factor which is soundly based is the rapid rise in real costs of LWR over the last five years. One study[4] suggests that the physical requirements of LWR have escalated as shown in *Table 10.2*. The major reason for these escalations appears to be increased safety provisions. As a result, cost estimates based upon

TABLE 10.1 CAPITAL COST ESTIMATES FOR LWR PLANTS (All costs are related to 1000 MWe plants and have been adjusted to mid-1977 prices.)

Source	Country	Capital cost (£/kWe)
Kunstige Stromgestehungkosten von Gross Kraftwerken, KFA, Julich, November 1977	Germany	360
Coal and Nuclear Power Station Costs, Energy Commission Paper No. 6, UK Department of Energy (1978)	UK	423
Nuclear Power—Issues and choices, Nuclear Energy Policy Study Group (1977)	USA	375
Coal and Nuclear Generating Costs, EPRI PS-455 SR, April 1977	USA	330–475
de Kosten van Kernenergie in Nederland, Koninklijk Instituut van Ingenieurs (1978)	Netherlands	475

apparently similar systems, and using the same cost date, may in fact refer to rather different reactor designs.

The work discussed here used 'high', 'medium', and 'low' capital costs figures of £525, £420, £315/kW, respectively, for direct nuclear capital costs, not including interest charges. In our view, the 'medium' figure must be taken as a bottom limit to the capital cost of any British unit. The LWR system has not been developed here to a full commercial size, whilst the AGR system is known to have unit capital costs which are at least 15% higher than the LWR. (All prices are mid-1977. Another 6-7% should be added to bring to mid-1978.) It will be seen that our 'medium' cost is slightly less than the UK Department of Energy estimate when this is adjusted to mid-1977 prices.

The capital costs of coal-fired plant used in recent studies are shown in *Table 10.3*. These costs have been adjusted to include full flue-gas desulphurisation (FGD) equipment. Present British environmental regulations do not require such controls, though they are standard in the USA and are likely to be required, for at least partial cleaning, in much of continental Europe.

It will be seen that these cost estimates are much more consistent. This does

122

TABLE 10.2 ESCALATION IN PHYSICAL REQUIREMENTS OF NUCLEAR AND COAL PLANTS[4]

Physical requirements	Nuclear			Coal		
	1974	1977	Change (%)	1974	1977	Change (%)
Building volume (ft³/kWe)	11	13	18	15	15	0
Structural steel (lb/kWe)	14	20	43	35	45	28
Reinforcing steel (lb/kWe)	28	39	39	7	8	14
Structural concrete (yd³/kWe)	0.11	0.14	27	0.06	0.07	17
Formwork (ft²/kWe)	1.5	1.9	27	0.5	0.6	20
Piping (lineal ft/kWe)	0.2	0.3	50	0.2	0.3	50
Conductor and cable (lineal ft/kWe)	2.7	3.7	37	3.2	3.5	9

TABLE 10.3 CAPITAL COST ESTIMATES OF COAL-FIRED PLANTS
(All costs have been adjusted to mid-1977 cost and include 100% flue-gas desulphurisation. Deduct £55/kWe for no FGD.)

Source	Country	Capital cost (£/kWe)	
Conceptual Design of Utility Steam Plant, General Electric, NASA CR-134950, December 1976	USA	313	
Coal-Fired Power Plant Capital Cost Estimates, Bechtel, EPRI AF-342, January 1977	USA	371	
Capital Cost: Low & High Sulphur Plants, United Engineers, NUREG-0244 (1977)	USA	309	
Coal and Nuclear Power Station Costs, Energy Commission Paper No. 6 (1978)	UK	349	Adjusted to include FGD
de Kosten van Kolenenergie in Nederland, Koninklijk Instituut van Ingenieurs (1978)	Netherlands	311	With regenerable scrubber
Kunstige Stromgestehungkosten von Gross Kraftwerken, KFA, Julich, November 1977	Germany	302	

not mean that coal-fired plant has not been subject to rapid cost escalation over the past few years. In particular, the increasing stringency of environmental controls has had a marked effect. The increase in physical requirements of coal plant, shown in *Table 10.2*, appears to be largely related to the US requirement for full flue-gas desulphurisation and the change in gas-scrubbing design. There does, however, seem less uncertainty over coal-plant costs, probably related to the fact that FGD units are an 'add on' to the main plant, in contrast to the LWR changes which have demanded complete reactor redesign. This is true both of planning estimates and also actual construction costs, where the limited information available suggests that coal-fired plant is less liable to cost overruns than nuclear units[5].

The study has used two coal capital cost estimates; one of £325/kW, corresponding to full desulphurisation controls, and £270/kW, corresponding to no sulphur removal. The implied estimates of £55/kW for flue-gas desulphurisation have been derived in detail in other EAS work[6]. (Again all figures refer to mid-1977 prices.)

A brief mention should be made of new coal technologies, in particular atmospheric fluidised bed (AFB), pressurised fluidised bed (PFB), and combined cycle gasification (CCG). All these are currently being studied as possible advances on the PF plant used for the cost estimates discussed already. The technical status of these systems is that they are operating, or plant is being built, at the pilot unit size, and that design studies are being made of scaling up to full commercial size[7].

The advantages claimed for these new technologies include:

(i)   lower capital costs;
(ii) greater fuel efficiency;
(iii) shorter construction periods;
(iv) small optimum unit size; and
(v)   better environmental control.

It is necessary to suspend judgement on these claims until full-size plant operation has been demonstrated, which is unlikely to occur before the end of the 1980s. Preliminary assessments suggest that the total cost of generating electricity may be reduced by 5–15% over PF systems, depending upon the cost of coal-feed and environmental controls. The main cost advantage of the new systems lies in their cheaper sulphur control system.

This suggests that the adoption of sulphur controls by the UK need not change materially the coal/nuclear balance provided these new coal technologies are developed in time.

10.3.2 Fuel Costs

The cost of coal is a basic parameter to any study of coal/nuclear electricity costs. The study reported here used high, medium and low coal costs of £1.58, £1.05, £0.53/GJ, ($3, $2, $1/GJ) as a reference coal cost. The current cost of British power-station coal, delivered to units close to a mine, is about £0.95/GJ. (The mid-1977 price which should, for consistency, be used to compare with mid-1977 capital costs, was about 10% lower than this.)

124

TABLE 10.4 MODEL MASS BALANCE AND PLANT FACTORS FOR LWR

(a) Plant factors

Plant capacity	1000 MWe
Plant efficiency	33.5%
Uranium burn-up	30 000 MWd/tonne
Enriched uranium	3% ^{235}U
Tail assay	0.2% ^{235}U
Waste uranium	0.85% ^{235}U
Load factor	65%

(b) Uranium flow

Stage		Uranium (kg per annum)	^{235}U (%)	SWU per annum*
1. Conversion	(in)	102 462	0.711	
	(out)	101 950	0.711	
2. Enrichment	(natural in)	101 950	0.711	
	(natural out)	18 606	3.0	81 478
	(recycled in)	22 636	0.85	
	(recycled out)	5 255	3.0	21 267
	(total out)	23 861	3.0	102 745
3. Uranium recycled from 4 and 5		1 761	3.0	
4. Preparation	(2% recycled, 0.5% loss)			
	(in)	25 622	3.0	
	(recycled)	512	3.0	
	(out)	24 981	3.0	
5. Fabrication	(5% recycle, 0.5% loss)			
	(in)	24 981	3.0	
	(recycle)	1 249	3.0	
	(out)	23 607	3.0	
6. Reactor	(in)	23 607	3.0	
	(out)	22 934	0.85	
7. Reprocessing	(1% loss)			
	(in)	22 934	0.85	
	(out)	22 705	0.85	
8. Reconversion	(0.3% loss)			
	(in)	22 705	0.85	
	(out)	22 636	0.85	
9. Waste disposal	(in)	902		

*The separative work unit (SWU) requirement is calculated on the basis of 4.306 SWU/kg of 3.0% enriched uranium. This is an average for the two feedstock streams.

Cont'd on page 126

TABLE 10.4 (cont'd)

(c) Annual operating requirements	
Mining and milling	120 831 kg
Conversion	102 462 kg
Enrichment	102 745 SWU*
Preparation and fabrication	25 622 kg
Reprocessing	22 943 kg
Reconversion	22 705 kg
Waste management	902 kg

*The separative work unit (SWU) requirement is calculated on the basis of 4.306 SWU/kg of 3.0% enriched uranium. This is an average for the two feedstock streams.

The fuel cost of a nuclear reactor is a much more complex matter. It requires, first, a fairly detailed model of the mass balances associated with any particular reactor, and secondly, a knowledge of the costs of each of the five or six steps involved in the fuel cycle.

The fuel model used in this report is shown in Table 10.4[8]. The fuel cycle of the AGR has a rather different balance but no detailed information is readily available. It has been reported[3] that AGR fuel costs are about 10% higher, per unit of electricity, than LWR costs.

Nuclear fuel has never fully entered the commercial market; many of the processes are intimately connected with plants originally built for weapons programmes or in the interest of national security, and there has been the same spirit of optimism pervading nuclear fuel estimates as capital costs. All these factors make cost estimates, particularly those based on historic data, vary widely. Table 10.5 shows a number of cost structures obtained from various sources and adjusted, where necessary, to fit the EAS fuel model. This also shows the values taken for the EAS calculations. These give a total fuel cost of £0.33/GJ or 0.35p/kWh. This corresponds very well with the estimate[3] of 0.36p/kWh, though in that study, it is assumed to be a 1985 cost.

The proportional breakdown of the fuel cycle shown in Table 10.5 is:

Mining and milling	33.4%
Conversion	1.3%
Enrichment	20.3%
Fuel fabrication	8.8%
Reprocessing	25.4%
Reconversion	0.6%
High-level waste storage	0.2%

It will be seen that fuel costs are dominated by three major costs. The appropriate questions to ask about future nuclear fuel cycle costs would therefore seem to be:

126

TABLE 10.5 FUEL CYCLE COSTS (all in £/specified unit)

Fuel cycle process	Source						
	Germany (1)	Netherlands	UK	Sweden	USA	Germany (2)	EAS model
Mining and milling (lb U_3O_8)	21	28	13	19	21–24	32	25
Conversion to UF_6 (kg)	2.5	2.5	n.a.	2.2	n.a.	3.1	2.5
Enrichment (SWU)	63	68	24	57	53–63	53	60
Fuel fabrication (kg)	88	73	64	64	58–63	126	70
Reprocessing (kg)	225	166	223	421	116–147	213	225
Reconversion (kg)	n.a.	8	n.a.	n.a.	n.a.	n.a.	5
High-level waste storage (kg)	n.a.	n.a.	n.a.	263	42–53	n.a.	50
Total unit fuel cost* (GJ)	0.33	0.34	0.21	0.37	0.25–0.30	0.37	0.33

*Based on unit requirements as shown in *Table 10.4.*

Sources
Germany (1): Private communication
Netherlands: *de Kosten van Kernenergie in Nederland,* Koninklijk Instituut van Ingenieurs (1978)
UK: Evidence to Windscale Inquiry reported in: Sweet, C., *Energy Policy,* June 1978
Sweden: Talmet, L. and Svensson, B., *Teknisck Tidskraft,* 16, 30–32 (1977)
USA: Delene, G., Private communication, Oak Ridge National Laboratory, May 1978
Germany (2): *Kunstige Stromgestehungkosten von Gross Kraftwerken,* KFA, Julich, November 1977

(i) How is supply and demand likely to affect future uranium prices? Some opinions suggest that prices may more than double to reach over £50/lb U_2O_3 by 2000, but these are based on a major expansion of nuclear power and fairly small increases in geologic reserves.

(ii) Have enrichment costs fully stabilised? The apparent price of enrichment has rocketed over the past few years, mainly it appears because of the removal of implicit government subsidy from enrichment plants. In the commercial environment of recently constructed plants, a more stable price may be expected but its sensitivity to capital cost increases and to electricity price rises has not been demonstrated. At least one Atomic Energy Authority source expects enrichment costs to rise in the future[21].

(iii) Will reprocessing costs fulfil, in practice, the technical studies on which quoted costs are largely based? The technology of reprocessing remains undemonstrated at commercial scale and must be considered liable to substantial cost overrun if past nuclear performance is considered.

10.3.3 Other Costs

Other costs, such as operating costs, working capital, etc., are a relatively small factor in the total costs of both coal and nuclear plants. The values used in the EAS exercise are:

	Nuclear	Coal
Annual operating costs	£5.26 million + 6% plant cost	£5.4 million at 65% load factor + £3.2 million + 5% plant cost
Working capital and start-up costs	£31.6 million	£22.6–25.0 million
Decommissioning	£26.3 million	–

The costs are rather above the 'works costs' quoted by the CEGB[22]. However, these CEGB costs appear to exclude important cost items such as insurance, certain kinds of maintenance, central office costs, and R & D. The total system costs, as contained in the full financial results, appear to match our figures quite closely, though a detailed breakdown is impossible.

10.3.4 Plant Factors

10.3.4.1 Construction Time

Power station construction periods vary according to a number of factors, including size, location, design and national construction productivity. The normal planned time for construction of a coal-fired power station is 6 years. Nuclear power stations take rather longer to build; a common planning assumption is a 7 year construction period.

There is evidence that actual nuclear construction periods are often longer than this and that construction periods are lengthening. Reasons for this may be that the larger reactor sizes require more on-site fabrication, that inspection may be more protracted and that hold-ups for licensing or other exterior controls may be lengthening. A report undertaken by the Mitre Corporation[5] shows that average construction time for US nuclear plant increased from 5.8 years for plants completed in 1971 to 7.8 years for plants completed in 1975. The construction time for coal plants remained steady over this period, though the number of plants considered was small.

British power plant construction periods are notoriously long, particularly in recent years, but it may be hoped that these will be reduced.

10.3.4.2 Plant Availability

There have been a number of surveys and analyses of large coal and nuclear generating plants[9,10,11,12]. These indicate that plant availability can vary widely from unit to unit with some stations performing almost without fault, whilst others are plagued with problems.

There is some evidence that large generating units perform less well than smaller, both for nuclear and fossil firing. A lot more data need to be considered, particularly for large nuclear units, but to date it seems unsafe to assume a lifetime load factor greater than 65% for either coal or nuclear plant.

This assumption ignores various complications about load factor which may be significant. In particular, it does not make any specific assumption about the variation of load factor over plant over a time. Generally, plant will be used at a higher load factor in its early years, provided that it is available. In later years, it may be expected that more recent plant may be preferred for base-load operation. (It is illegitimate to assume that this will be a preferential factor for nuclear units, as in one recent study[3], as this involves prejudging the results of cost comparison.) It is assumed below that nuclear units have a longer start-up period, which corresponds to assuming a lower load factor in early years.

10.3.4.3 Start-up

It is normal for all large plants to operate below capacity in their first period of operation. The actual output build-up attained by any particular plant is very variable. Some plants may operate smoothly from the first months of operation, others may be plagued with recurrent faults for years. We have assumed a some-what extended output build-up for nuclear plants as compared with coal. The figures used are 30% in year 1.50% in year 2, reaching 65% in year 3 for nuclear, as against 32.5%, 65% for coal.

This protracted build-up relates partially to the greater complexity of nuclear systems and the greater difficulties of remedying reactor faults, and partly to the requirement for safety inspection and licensing, and the other regulatory delays to which nuclear, but not coal, are subject in their first years.

10.3.4.4 Shut-down

The closure of a nuclear plant poses some particular problems associated with the radioactivity of the reactor and possibly some components in the steam circuit. Three levels of action have been discussed with respect to the treatment of shut-down reactors.

The first essentially involves the sealing of the reactor and some kind of permanent security arrangements to protect against human access. The second involves dismantling major parts, after some period of time, and the permanent entombing of the remainder. The third level involves the complete dismantling of all parts of the reactor. It is not known whether this third stage is feasible in practice.

The cost of any of these operations and the actual level of action which may be required is not known. All reactors which have been shut down to date have been small units within nuclear establishments. A figure of £26.3 million has been added as a final-year cost to nuclear power stations as a contingency for these, largely unknown, costs[14]. The actual size of this cost is, in any case, largely irrelevant in any discounting calculation.

129

10.3.4.5 Plant Life

Fossil-fired steam power plants are known to have a lifetime of up to 30 years or more, though it is normal for the older units to be used with a much lower load factor. The lifetime of nuclear power reactors remains unknown, as no large unit has operated for its full life cycle. A plant lifetime of 25 years has been assumed for both types of plant.

10.4 GENERATING COSTS

The results of this study can be presented in a great variety of different forms. The figures given here attempt to answer three questions about the coal/nuclear cost comparison.

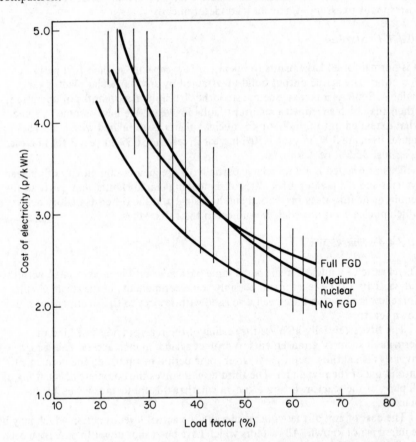

Figure 10.1 Costs of coal and nuclear electricity. Assumptions: 10% rate of interest; coal price of £0.95/GJ; nuclear fuel cost of £0.33/GJ; no inflation; nuclear capital costs, £525/kW−high, £420/kW−medium, £315/kW−low.

What is the competitive position, at current prices, over the full load range? What is the change in competitive position at different discount rates and with new technology? What is the effect of real cost inflation of fuel price?

Figure 10.1 summarises the competitive position, at constant current prices, for current British coal price of £0.95/GJ against high, medium and low nuclear capital costs for coal units with and without flue-gas desulphurisation (FGD).

If the medium nuclear case is taken as the base for comparison, then *Figure 10.1* shows that, for coal plants without flue-gas desulphurisation, coal-based electricity is cheaper than nuclear at coal prices below £1.20/GJ. The requirement to use 100% FGD lowers this break-even coal price to £0.81/GJ.

Figure 10.2 summarises the electricity costs of new coal technologies, in particular AFB and PFB, as compared with conventional coal (PF) use, at discount rates of 5, 10, 15%. This figure is calculated at a 65% load factor. It can be seen that AFB and PFB have no real advantage over PF boilers if no sulphur control requirement exists. Their cost advantage lies in the low cost methods of sulphur control which these systems allow.

The use of various discount rates shows, as expected, that at the lower rate,

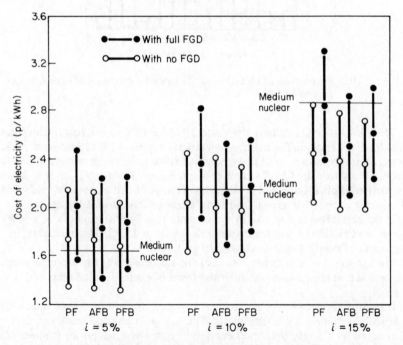

Figure 10.2 Electricity costs for coal and nuclear plants at different interest rates. Uppermost dot gives cost at coal price of £1.60/GJ; middle dot at £1.00/GJ; lower dot at £0.50/GJ. PF, pulverised fuel; AFB, atmospheric fluidised bed; PFB, pressurised fluidised bed.

131

nuclear power has a greater advantage over coal because of the relatively high capital cost component. The break-even coal cost for PF firing without sulphur control is £0.92/GJ at 5% discount rate and £1.63/GJ at 15% discount rate. This all relates to LWR units, which are, as yet, not built in the UK. General information suggests that these break-even prices could increase by about 10% if the AGR were to be used as the standard of comparison.

These results suggest that, at current prices, nuclear power has no competitive advantage over coal, for base-load generation, if current environmental controls continue to apply. The cost margin for coal is significant at a 10% discount rate though small at a 5% discount rate. The cost advantage of coal decreases with any requirement for sulphur emission control, though new coal combustion technology will alleviate this.

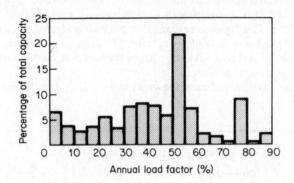

Figure 10.3 Proportion of CEGB capacity operating at various annual load factors in 1977–78[22].

These break-even costs have been calculated for 65% annual load factor, that is 'base-load' generation. The nature of electricity supply, with little storage capacity, means that only a fraction of power is generated by plants operating at base load. *Figure 10.3* shows the CEGB system load characteristics in 1977–78, that is, what proportion of plant generated at what load factors. It will be seen that only 16% of capacity operated at annual load factors above 60%. At load factors below 60%, the cost advantage to coal increases quite rapidly. For example, at a 50% annual load factor the coal break-even price is about £1.12/GJ, an increase, in real terms, of nearly a quarter above current levels.

The economic argument between coal and nuclear power therefore turns upon future real cost changes, in particular the trend of nuclear capital costs and coal costs.

It has already been noted that the costs of nuclear plant have been increasing at well above the general inflation rate. The change in physical needs has already been noted for US reactors. Another study[15] notes that manpower requirements for construction of LWR rose from around 5 man hours/kW for plants starting up in the early 1970s to the 10 man hours/kW estimated for a plant due to start up in 1983. In Britain, the cost escalation of the AGR programme is well known: Hinkley Point B is reported[3] to have gone up by 30% in real terms whilst the implicit escalations of the uncompleted reactors is certainly rather higher. The new round of AGRs, beginning with Torness, appear to involve substantial re-design. It would be unusual if this did not result in even higher costs.

The problem is, at what level will the real cost of nuclear generating plant stabilise? Optimism may suggest the conclusions of the official study that 'changes in the last few years in industrial structure and in our approach to research and design and development . . . should improve prospects for the future'[3]. But past cost escalations of the AGRs (which already show a 15% capital cost differential over the LWR costs used here), and the lack of any substantial British capability in LWR design and construction may suggest a harsher judgement.

It would also, perhaps, be optimistic to suggest that British coal prices will remain constant in real terms, and indeed it has become clear that the economic advantage of nuclear power lies in making some assumption about future real cost price increases for coal. This view is confirmed, albeit negatively, from the most unlikely source. 'Unless coal prices remain somewhere near their present level in real terms over the next 25 years—surely an unlikely event particularly in the absence of nuclear power—or nuclear capital costs escalate much more than fossil-station capital costs, there will still be a large economic advantage in instal-ling nuclear rather than coal-fired stations'[23]. Just so. The question remains as to just what combination of real cost rises can occur for coal to retain its competi-tive position.

The work conducted at EAS has considered four coal/nuclear capital cost structures in which coal has either no FGD or full FGD and in which nuclear costs stabilise either at current values of around £420/kW or at a level 25% higher. The maximum real cost escalation which could be suffered by coal, starting from current British prices, and remain competitive with nuclear power at 65% load factor, is shown below:

100% FGD/medium nuclear	not competitive
100% FGD/high nuclear	2.2%
No FGD/medium nuclear	2.5%
No FGD/high nuclear	5.3%

In all cases, it is assumed that nuclear fuel costs escalate at the same rate as coal.

The likely long-term trend in UK coal prices is not something which can be adequately discussed here. However, two points can be made. First, it is not easy to sort out any useful trend on past behaviour; too much depends on choice of comparison years. (In 1969, the real price index of coal was 87.7; in 1972, it was 115.4; in 1974, 66.6; in 1976, 89.3[24].) Secondly, the major coal investment pro-gramme at present being undertaken is posited on major productivity increases in coal mining. This must act to stabilise the real price of coal notwithstanding contrary trends such as the real value of miners' wages.

We may note that the conclusions reached above are, in general, shared by the official British analysis quoted[3], though mainly for methodological reasons they derive costs more favourable to nuclear power. They conclude that, at real cost rises of something below 1%, coal generating costs are superior to all reactor systems considered, with capital cost assumptions equivalent to our no FGD/

medium nuclear cost. The gloss put on their discussion favouring nuclear power derives from a base-case scenario which assumes a real price rise of 4% per annum for coal over the next 20 years.

10.5   THE FAST BREEDER REACTOR: GENERAL ISSUES

The preceding sections may, in the context of this volume, be regarded as only a preamble to the issue of establishing the economics of the FBR. The preamble will be seen to be rather larger and more precise than the main work, for reasons discussed below. However, two important factors have been derived.

First, the costs of coal-based electricity have been derived, with reasonable confidence, apart from establishing the future path of the real price of coal. This is important as coal plant is the medium-term competitive alternative to the FBR and therefore sets the economic standards against which the FBR must perform.

Secondly, the costs of thermal-reactor electricity have been derived, though with rather less confidence about the current capital costs of reactor units. This is important because most of the published work on the economics of the FBR uses LWR costs as a yardstick.

There are two, rather general, comments which can be made about the economics of the FBR. First, it *is* required to have an economic justification in addition to the general justification of depending upon a large natural resource base. It is a comparatively simple matter to conceive of generating systems, base load or otherwise, relying on unlimited natural resources. (Wind generators electrolysing water to produce hydrogen fuel for gas turbines is an obvious and technically simple example.) The capital cost of these would normally be very large and they are unlikely to be considered at any likely coal price. However, the point is that such systems do exist and the FBR has no *a priori* advantage over them.

Having said that, it must be admitted that the FBR, as a commercial project, does not yet have much economics, in the sense that no proper techno-economic evaluations appear to have been done; certainly they have not been published. This is not the least odd feature of FBR development. It is easy enough in the field of energy economics, to be overwhelmed by the sheer volume of technical evaluations and economic feasibility studies which pour forth on every likely and unlikely energy source. Coal-gasification studies or solar-heating evaluations, for example, could easily form a fuel source in their own right. The FBR has, however, attracted much less attention; it is a restful haven in the hurly-burly life of an energy economist.

This, it must be suggested, is a rather unusual situation for a system which is now being built at full commercial size in France and already exists at about one-quarter full size in the UK. However, even lacking much more than fragmentary information, a discussion of FBR economics can proceed at a number of levels.

The first, that of national energy security, to some extent passes out of the realm of economics though, as noted above, this does not obviate the need for an economic contribution to the discussion. There is, underlying much of the discussion on the FBR, a strong element of national energy autarky, something which is very apparent and openly expressed in the French programme. This can enable costs to be relegated to a secondary position.

In Britain, the situation with regard to such a goal of energy self-sufficiency is complicated both by a reluctance to adhere, at least openly, to such an aim and by the fact that fossil-fuel reserves are sufficient in principle to achieve substantial independence for many decades. The policy issue is a matter of political judgement and will not be pursued here, but the second point has the effect of introducing economic choice even into a debate centred around energy autarky. If both coal reserves *and* depleted uranium stocks can allow some hundreds of years supply, not to mention regenerable energy, then the problem of *choice* immediately becomes paramount.

Another level of debate, what may be called the ethico-economic, is an evident strand to the FBR debate, concerned with the relationship between energy supply and economic growth. On the one hand, there is maintained the presumption that high economic growth rates are tied to equally high, or even higher, rates of growth in energy demand and that maintaining a non-resource-limited energy supply is a vital factor in ensuring future economic prosperity. On the other hand, there is the position which suggests that this automatic linkage is invalid, based upon the too-ready transference of a particular, and transient, stage of historical development to the status of immutable law. I have called this an ethico-economic debate not because there is not an enormous amount of technical data to be flung back and forth in support of these views, but because it is difficult to escape the idea that the two positions contain underlying social and ethical assumptions about a desired future path of society. As such, it is entirely possible for both positions to be self-fulfilling in terms of policy actions.

Paradoxically, whereas the first issue, that of energy security, very rapidly assumes an economic face in the British context, the second, which seems at first sight to be bound up in the technical minutiae of energy usage, very quickly moves away from the economic and into other, more diffuse, areas.

A final general point is that concerned with national technological superiority. Again this is a difficult issue to quantify, but it is evident that the FBR carries with it, almost as a sole survivor, the belief that technological advance in certain fields—of which nuclear power is one and, say, mining machinery is not one— carries with it a national clout that cannot be expressed in direct financial benefit. Perhaps the clearest expression of this is the idea that some centre of technological prowess is required to attract, inspire and motivate engineers and scientists throughout society.

Whatever its justification, and whether or not it is true, the factor of technological advance as a virtue in its own right cannot be wholly excluded from any discussion about the FBR.

All this may be taken as general background to the issue of the precise economics of the FBR. The point of taking up some space to discuss such matters is to emphasise that, whatever answers come out of the sums, actual political decisions are shaped by wider perspectives.

10.6 THE ECONOMICS OF FAST BREEDER REACTORS

The economics of the FBR are a good deal more complex than the simple, single station comparisons discussed above. There are a number of questions which can only be properly answered by an extensive analysis of the whole energy economy of the UK over the next 50 years or longer. It is the need to answer such questions which gives the FBR project such a strategic place in energy policy. Much more than any single decision about coal or thermal nuclear stations, the projected FBR programme contains implicit or explicit assumptions about very wide issues of energy use and supply.

This strategic importance of the FBR derives essentially from the critical role played by various growth rates, or rates of change, in any economic justification of the FBR. Amongst these can be included:

(i) rate of growth of energy demanded;
(ii) rate of growth of electricity usage as a component of that demand;
(iii) rate of growth of thermal nuclear component in electricity supply;
(iv) rate of change of pattern of electrical demand; and
(v) rate of change in both supply and price of fossil fuels.

All of these, it will be noted, lie outside any characterisation of the techno-economic parameters of the FBR itself, though these determine the exact relationship between the FBR programme and the wider issues summarised above. The reason why such a close linkage exists between the general rates of change of the energy economy is that the FBR programme can only be justified at above some minimum growth rate for the building of FBR electricity. Just what is this growth rate remains an unresolved issue, but a key feature of the analysis is that some minimum growth rate does exist.

Its existence falls out of an examination of the single station economics of the FBR, a topic which will be briefly examined.

TABLE 10.6 HISTORIC FBR CAPITAL COST ESTIMATES ($/kW, 1974 dollars)

Power plant type	Date of introduction					
	1974	1980	1985	1990	2000	2020
LWR (1300 MW)	324	324	324			
LWR (2000 MW)				284	284	284
FBR (1300 MW)			402	376		
FBR (2000 MW)				339	284	284
Coal (1300 MW)	270	270	270	270		
Coal (2000 MW)				246	246	246

(All costs have interest during construction included.)

The most comprehesnive published attempt to quantify FBR costs is that provided by the US report on the commercial FBR project[16]. The power plant capital costs projected in this study are shown in *Table 10.6*. These estimates are, now, only of historic interest, but they demonstrate one essential element of FBR cost estimates, that of applying a learning curve to its development, such that, in this case, 35 years after its initial introduction, FBR unit capital costs are predicted to decrease by nearly 30% in real terms.

It would be easy to criticise these assessments, particularly such statements as 'the LWR, in the opinion of [Atomic Energy Commission], has reached a mature stage and further overall plant cost increases (in terms of 1974 dollars) will not be large'[17] in view of figures such as *Table 10.2*. What is relevant here, however, are two features:

(i) that initial studies were based on the assessment that FBR capital costs would eventually equal thermal reactor costs; and
(ii) that a substantial drop in capital costs over the first units was required to provide a satisfactory economic rationale for the FBR.

These assumptions were shared by the French, at this stage of FBR development. One quoted comment is: 'Results of detailed study show that, from the first plant to the sixth one, a cost decline may be expected of 20 to 30%, which gives FF1000 to 1150/kW ($235-270/kW)'[18].

The prediction of such a rapid fall in unit costs has been criticised from a number of quarters. One of the most telling derives from the President of the Rand Corporation[19], who suggests that the capital cost differentials are 'highly unrealistic', and that any likely learning curve applied to the capital costs at which the FBR will be introduced ($2000/kW in 1974 dollars) means that FBR 'technology that is not competitive with either LWR or fossil-fuel generators is thereby implied'.

The initial (if 1974 can be termed 'initial') optimism has clearly been confounded. In particular, it now seems doubtful whether FBR will ever have unit costs comparable to thermal reactors. This change appears to derive from a changing attitude towards FBR design. In 1974, for example, the AEC could confidently assert that no secondary heat exchanger was required between sodium and steam circuits. By 1978, this requirement appeared to be standard. Current economic justification for the FBR is now based upon future fuel cost escalation (nuclear as well as fossil) as well as predictions of capital cost decreases in FBR plants:

'Creys-Malville [the French, 1240 MW, Superphenix FBR] is certainly more expensive than an LWR of the same power—about double. The fabrication and reprocessing of its fuel in a proper fashion will also be more expensive. But the cost of a prototype is certainly not comparable with the construction of a series of a proved model. However, FBR stations will probably remain more expensive than building LWR units, but this should be compensated for by a lower cost of fuel that does not need enrichment of natural uranium, the price of which will inevitably rise with increasing world consumption'[20].

'Because thermal reactors will have to bear prices for uranium and enrichment adequate to encourage expansion of supply, fast reactors could cost more than thermal reactors and remain competitive. However, the allowable margin is limited, and early fast reactors are likely to exceed it. But with further development based on manufacturing and operating experience, construction costs should be brought within the required margin.

Also the prices of uranium and separative work are expected to increase as prices of competing fossil fuels go up'[21].

It is helpful at this point to attach a few numbers to these statements. Such numbers can only be tentative but they are probably not unreasonable.

Let us assume that the FBR will enter commercial use at £1050/kW, following the judgement of the Rand Corporation[19], and that its eventual costs stabilise at £525/kW, the high nuclear estimate used above. The fuel cost of the FBR is not known, but it would be optimistic to assume a figure of £0.18/GJ or 55% of our calculated thermal reactor fuel cost. (This is assuming that a programme of sufficient size to justify full-scale fuel plant exists. Fuel costs would be higher in initial phases.)

Electricity costs of these plants can be then calculated at about 4.8p/kWh and 2.6p/kWh; (this assumes similar plant factors to the LWR), as compared with current costs of about 2.5p/kWh for LWR and 2.05p/kWh for coal plant without FGD (*Figure 10.1*).

In the case of LWR, although this cost differential is quite small, it would require a fuel price rise from £0.33/GJ to £0.59/GJ or 1.8% per annum from now until 2010, to make the FBR competitive even at its suggested final cost. This is clearly not an impossible rise, though nearly double that used by the Department of Energy in their recent study[3], which proposed an LWR fuel price rise from 0.36p/kWh in 1985 (about £0.30/GJ) to 0.461p/kWh in 2010; an annual growth of just under 1.0% per annum. At this growth rate, FBR electricity would be competitive with LWR in 2036 provided that the lower capital cost level had been attained.

It is also clear that fairly substantial rises in the real price of coal are required to make FBR electricity competitive, even if the substantial capital cost decreases used in this calculation are achieved.

The other part of the scissors, the falling unit FBR cost, which is to meet rising fuel costs can also be examined using loose, but not unreasonable, numbers.

The US study noted above[16], envisaged a massive FBR programme in which the first commercial unit would be introduced in 1986/7 rising to 992 plants being built between 2000 and 2019. By 2020, it was supposed that 1178 plants would be operating. It will be remembered from *Table 10.6* that this scale of programme would lower real unit costs by 29%, including both learning effects and the scale advantages involved in moving from a 1300 MW size to 2000 MW. This latter gave a 9% reduction, so the effect of learning on unit capital costs was assumed to be 20%.

The scale of this proposed expansion is significant, even though the actual costs used are no longer relevant, for it indicates the US belief that rather large series production of the FBR would be required to make it economic, even at the small capital differential of 24% over LWR.

The learning curve used in this case was 98% (in learning curve theory this means a 2% reduction in capital costs for each doubling of unit production). If a 95% learning curve is used, then the reduction in costs from £1050/kW to £525/kW would require the building of over 11 000 units; if a 90% curve, then about 95 units would be required. The French report quoted above[18], claims a 90% learning curve is possible for FBR. Recent French figures tend to be less optimistic about FBR costs in general, but let us accept the 90% figure as being a 'best case' scenario.

It is at this point that the techno-economic argument specific to the FBR collides with the wider energy issues noted above, for, as is well known, the pace at which breeders can be built depends not upon their own technical specifications but upon the growth rate of thermal reactors.

138

The growth rate of the FBR programme which can be self-sustained is the inverse of the linear doubling time, that is, the time taken for an FBR to produce sufficient plutonium to provide the initial fuel inventory for another FBR. Recent estimates suggest that, on introduction, this doubling time will be about 50 years[21]. This reference touches on the complex problems of optimising FBR design with respect to this and other parameters, such as fuel rating and capital cost. It is suggested that 50 years may be reduced to 25–30 years in the course of a programme, but it is not clear what has to be sacrificed to achieve this figure nor why 50 years is regarded as an initial optimum.

Taking a 40 year doubling time as an average, and accepting that 95 units are required before capital cost reductions make the FBR economic, then 185 years would pass before a self-sustaining programme could achieve such cost reductions. Of course, this figure has no base in reality because it is assumed that an FBR programme would be fuelled in its initial stages by thermal reactors, whose growth rate is unconstrained by plutonium supplies. What then becomes important is the rate at which thermal reactors are built, and the existing stocks of plutonium and unreprocessed fuel from which plutonium can be extracted.

The latter is a figure shrouded by security, but it is unlikely to be significant over and above military requirements. To be simple (and there is no reason to be complicated when playing with such vague figures) let us assume existing plutonium sufficient to fuel an initial FBR and a thermal reactor capacity of 16 GW in 1990, when this first unit comes on-stream. Such capacity would provide plutonium for 1 FBR annually (at a production rate of 250 kg Pu/per annum per GW and an initial 1 GW FBR fuel inventory of 4.0 kg Pu per MW).

If thermal reactors were, thereafter, built with a growth rate in capacity of 10% per annum, then 95 FBR units could be built after 38 years; if thermal reactor capacity grows at 3%, then it will take 85 years. The former case assumes a total nuclear capacity in 2020 of 313 GW; the latter gives a figure of 44 GW. This means that, on the simplified assumptions used here, the FBR would achieve a steady capital cost sufficient to make it competitive with thermal reactors in 2028 in a high nuclear growth case and 2075 on a low nuclear growth case. It hardly needs demonstrating that high or low nuclear growth is itself a derived factor, depending upon nuclear and fossil-fuel costs and the general level of demand for energy and, in particular, electricity. The horizon for FBR viability can thus be seen to be very sensitive to assumptions about the rate of increase of thermal reactors, and therefore to rates of change throughout the UK energy economy.

Such calculating exercises very rapidly become a demonstration of the runaway properties of exponential growth, so it is possibly better here to summarise the discussion rather than press yet further into a murky future.

10.7 CONCLUSIONS AND QUESTIONS

It is useful to divide this summary into conclusions, that is, matters which seem to be fairly clearly established, and questions about matters which remain unclear.

Conclusion 1: The main competitors for electricity generation are coal and nuclear power. At current prices, coal-fired stations have a clear competitive

139

advantage over all thermal reactor systems. All conclusions as to the competitiveness of nuclear systems rest upon assumptions as to future real cost movements.

What are the likely future costs of UK coal and capital costs of nuclear power stations? In the past, both have shown a marked tendency to rise. Will the massive investment in new collieries stabilise real coal costs? Is there any sign of the British nuclear power industry managing to stabilise the real capital costs of nuclear power plants? To what extent are the investment programmes in coal and nuclear independent, and to what extent does the economic justification of one depend implicitly upon the failure of the other?

Conclusion 2: The FBR will be introduced at a capital cost which will make it uneconomic with respect to both thermal nuclear and coal at almost any conceivable fuel price. Its economic justification depends upon reduction of capital cost to a level at which it can compete with possible fuel price increases.

Capital cost reductions are normally analysed using learning curve theory. Is there any substantial justification for this as the theory was developed essentially for production runs over much shorter periods than is usual for large plant investment? What is the basis for real cost decreases claimed to have been associated with some LWR programmes?

Conclusion 3: The most optimistic assessment of capital cost reduction is a 90% learning curve. This requires that 95 units be built to achieve the capital cost reductions of conclusion 2 using reasonable values for initial and final capital cost. A 95% learning curve would require the construction of over 11 000 units. The construction of even 95 units would require a time scale of 185 years if an FBR programme were to be self-sustaining. To bring this time scale down to a reasonable period requires the use of plutonium derived from an expanding thermal nuclear programme. General estimates suggest that a 10% thermal nuclear growth rate could reduce the time to 38 years; a 3% growth rate to 85 years.

A corollary to conclusion 3 is that an FBR programme can only be justified in a reasonable time scale at electricity demand growth rates rather in excess of commonly accepted general energy and economic growth rates. This implies that energy supply should be increasingly based on electricity and that a larger and larger proportion of national income should be invested in electricity generating facilities. Is this conclusion justified in terms of likely price differentials with other fuels? Is there any upper limit to the number of thermal nuclear reactors which can be built in the light of likely uranium supply, and does this limit conflict with the growth requirements to fuel an FBR programme? Do the implications for investment in electricity generation units have any interaction with general economic growth? Is electricity investment at such rates likely to crowd out other investment in, say, manufacturing industry? All available information suggests that increases in the real cost of electricity from *any* source are inevitable. What interaction is likely between this and demand for electricity? What is the minimum electricity growth rate required to justify an FBR programme?

140

Conclusion 4: The judgements made in assessing the FBR programme go way beyond simple single-station techno-economic comparisons and involve very important assumptions about energy growth and economic development.

Have we, or are we likely to have, enough information to tackle the problem with any chance of making the right choice? What socially acceptable procedures can we adopt to ensure that *whatever* choice is made will be accepted by, at least, a substantial part of all social interests concerned?

REFERENCES

1. The details of this procedure have been reported in a series of EAS reports on *Technical and Economic Criteria for Coal Utilisation Plant*
2. *The Choice of Thermal Reactor Systems*, National Nuclear Corp. Ltd (no date given)
3. *Coal and Nuclear Power Station Costs*, Energy Commission Paper No. 6 (1978)
4. Bowers, H. I., 'Capital Investment Cost Estimates for Large Nuclear and Coal-Fired Power Plants', *Seminar on Electric Utility Generating Costs in 1980's*, Washington, D.C., 29 June 1977
5. Mitre Corporation, *Analysis of Project vs. Actual Costs for Nuclear and Coal-Fired Power Plants*, FE-2453/2, September 1976
6. Prior, M., 'The Control of Sulphur Oxides Emitted in Coal Combustion', *EAS Report* B1/77, December 1977
7. A full evaluation of the economics and technical status of advanced coal-based electrical generating systems is being undertaken by EAS and will shortly be published
8. This model is based on that contained in: Rieber, M. and Halcrow, R., *Nuclear Power to 1985*, Centre for Advanced Computation, University of Illinois, CAC No. 163, November 1974
9. Howles, L. R., 'Review of Nuclear Power Station Achievement–1976', *Nuclear Engineering International*, April 1977
10. *US Central Station Nuclear Power Plant: Operating History 1976*, ERDA 77-125 (1977)
11. *Availability of Fossil-Fired Steam Power Plants*, EPRI FP-422 (1978)
12. *Steam Electric Plant Factors*, US National Coal Association (1977)
13. *Electricity Supply Handbook*, Electrical Times, London
14. A discussion of this cost is contained in: *Disposal of Nuclear Radioactive Wastes*, OECD, November 1977
15. *Nuclear Power: Issues and Choices*, Nuclear Energy Study Policy Group, Ballinger, Cambridge, Mass., 1977
16. *Final Generic Environmental Impact Statement on LMFBR*, WASH 1535 (1975)
17. *Final Generic Environmental Impact Statement on LMFBR*, WASH 1535, Vol IV, p. 11.2-86 (1975)
18. *Final Generic Environmental Impact Statement on LMFBR*, WASH 1535, p. 11.2–91 (1975) (quoted therein)
19. Alexander, A. J. and Rice, D. B., *Comments on LMFBR Cost–Benefit Analysis*, Rand Corporation, AD-AO22-296, August 1975
20. Banal, M. and Megy, J., 'Super-Phenix, Premiere etape du development commercial des surreegenerateurs', *Annales des Mines*, Mai-Juin 1978

21. Farmer, A. A. and Hunt, H., 'Important factors in fast reactors' economics', *Nuclear Engineering International*, July 1978
22. *CEGB Statistical Year book, 1977–78*, CEGB, London (1978)
23. Hunt, H., 'Nuclear Power Costs in the UK', *Energy Policy*, December 1978
24. *Handbook of Electricity Supply Statistics, 1977*, Electricity Council, London (1977)

11

Nuclear Power Economics

P. M. S. Jones

11.1 INTRODUCTION

I intend to devote most of my paper to the presentation of some elementary points. For discussion of the concepts, it matters little whether we compare nuclear and fossil energy sources, thermal and fast reactors or two fossil sources. However, for purposes of illustration, I will stick to the comparison of advanced gas-cooled reactors (AGRs) and coal-fired power stations in the first instance, and introduce fast reactors later to deal with their similarities and differences.

11.2 POWER STATION INVESTMENT DECISIONS

11.2.1 The Simple Economic Comparison

The simplest case to consider is one in which there is a fixed steady demand for electricity which can be met by one of the two power stations under consideration. We then have the choice of either an AGR with high capital cost but low operating and fuel costs, or a coal-fired station with lower capital cost and high fuel costs.

If we know the life of the two stations, can assume equal operational availability, and have reliable estimates for their capital costs and their future operational and fuel costs, then we can make a comparison. It should be pointed out, however, that there is no unique capital cost, it is dependent on the specific design built to meet operational specifications and to conform to environmental and safety regulations and requirements laid down by the Government or the plant operator. For this reason, the capital costs for the power station, of whatever type, may well be different in the United Kingdom from the costs of a broadly equivalent station for construction in the United States or mainland Europe. The differences will arise not only from the factors mentioned above but also from differences in material and labour costs in the two countries which are not necessarily in line with the monetary exchange rate.

For these reasons, the capital costs appropriate for comparisons in the UK are those estimated on the basis of designs for construction in the UK. The most up-to-date published figures on which to base our comparison are those supplied for

143

nuclear reactors to the Secretary of State for Energy by the National Nuclear Corporation[1] and the CEGB[2]. Other figures have recently been announced by the CEGB[3] for Drax 'B'*. The former figures were used in Energy Commission Paper No. 6[3]. The NNC and earlier CEGB figures are used in *Table 11.1* in terms of £/kW present worth at 1 January 1977 prices. These figures assume a 25-year life for the nuclear station and a 30-year life for the coal station, with both operating at 70% load factor. They relate to new stations scheduled for commissioning in 1985 and are calculated on the basis of the required rate of return specified for public sector investment of 5%[4]. An allowance is also included in the table for interest during construction.

As indicated above, the comparison of the two stations has to employ some value for the future cost of the fuels and for other operating costs. Coal has been taken to cost £30/ton in 1985 (1978 money), compared with the current average £25/ton for power station coal in the UK. This 20% rise in real terms over 8 years does not appear unlikely in view of the greatly increased levels of investment and the trend in wages. Uranium has been assumed to cost $40/lb U_3O_8 (1978 money) in 1985. This is the price at which new long-term contracts are being let compared with existing contract prices of about $20/lb. *Table 11.1* includes the costs of the initial fuel for working stock, a small allowance for the value of the residual fuel

TABLE 11.1 COMPARISON OF AGR AND COAL-FIRED STATION COSTS†

| | Present worth at 70% load factor (1 January 1977 prices) (£/kW) | | | |
	AGR		*Coal*	
Construction cost (settled design)	470		290	
Interest during construction	70		44	
Total station cost	540		334	
Initial fuel or working stock	68		7	
Final fuel	4		–	
Fixed operation costs	76		55	
Total fixed cost		688		396
Replacement fuel	361		1183	
Variable operating costs	38		27	
Total running cost		399		1210
Generating cost (£/kW) (rounded)		1100		1600
Generating cost (p/kWh)		1.28‡		1.70

†For commissioning in 1985.
‡This figure includes an allowance of 0.05p/kWh to allow for decommissioning and waste-disposal costs.

*Drax 'B' has 3 × 660 MW units (net output 3 × 625 MW) and will cost £685 million excluding interest during construction. In 1978 prices, this is equivalent to £365/kW compared with the £290/kW given on January 1978 in Energy Commission Paper No. 6.

in the nuclear reactor at the end of its life, and allowances for fixed operating costs and for decommissioning of the nuclear reactor. The replacement fuel charge for the nuclear system includes the costs of reprocessing and an additional sum has been incorporated for the ultimate disposal of the radioactive wastes. Decommissioning and waste-disposal charges have not been included for the fossil station.

The figures in *Table 11.1* differ from those given by the National Nuclear Corporation[1] in minor detail and in that they employ the recommended 5% discount rate (net of inflation) rather than the 10% employed by NNC. The figures also differ from those recently tabulated by Hunt and Betteridge[5] since they

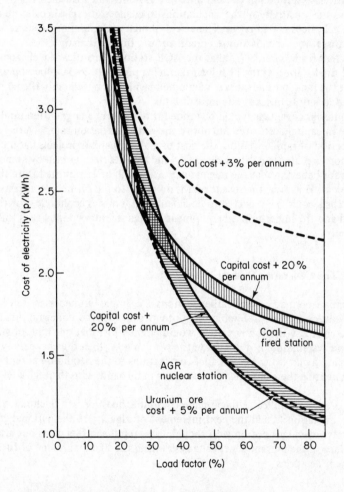

Figure 11.1 Comparative generating costs for coal and nuclear stations in 1985. Discount rate of 5% per annum. Lifetimes of 30 years for coal and 25 years for AGR. Capital costs: coal, £290/kW + 15% IDC; AGR, £470/kW + 15% IDC. Coal cost of £30/ton + 5% handling charge; uranium ore cost of $40/lb U_3O_8.

incorporate the small allowances for decommissioning and waste disposal covered by the cited authors in a footnote.

11.2.2 Sensitivity

It will be evident that any estimate of capital, fuel or operating costs which has to look a significant period into the future is likely to be subject to error. *Figure 11.1*, taken from Hunt and Betteridge[5], compares the costs of the AGR and coal-fired stations as a function of discounted average lifetime load factor and illustrates the sensitivity of the calculations to changes in capital costs and inflation of fuel prices in *real terms*. The effect of higher uranium ore costs at low load factors arises from the discounted credit for the final fuel charge.

It is clear from *Figure 11.1* that the AGR station is significantly cheaper than the coal-fired station at the high load factors appropriate to base-load operation and that the margin of advantage would be magnified considerably if coal prices increased at any significant rate in real terms.

The nuclear advantage would be eroded if there were a large differential inflation in capital costs over and above the general inflation rate.* During the recent period of rapid inflation, the cost of all large capital projects has increased much more rapidly than prices generally, and this has been argued by some as grounds for believing that the nuclear advantage will be less in the future than it is at present. However, the recent past is no guide to the future in this respect. Capital costs could be expected to decline in real terms as economic growth is resumed and the under-utilised equipment in manufacturing industry is fully employed.

11.2.3 The Generating System

The comparison made in the earlier sections is a simple but necessary one. If one type of station proved preferable to all others over the whole range of load factors, then the cheapest generating system would employ just that one kind of station. When one system does not dominate the whole range, then the cheapest system will employ a mix using the low-fuel-cost stations to the maximum extent possible and augmenting their output, to meet peaks in demand, with the high-fuel-cost stations.

This greatly complicates the apparently simple question of the choice of the best power station to add the next increment of capacity to the national grid. The decision is dependent not only on the relative capital and fuel costs but also on the balance in the present generating system and the likely outcome of future investment decisions.

---

*It is conventional in the UK to quote all costs in constant money terms and these are unaffected by inflation provided the costs of labour, materials, etc., rise in step. It is possible for different sectors of the economy to have different rates of price rise and it may be appropriate to take such differential inflation into account in constant money calculations using an inflator or deflator to take account of deviations from the mean, as appropriate.

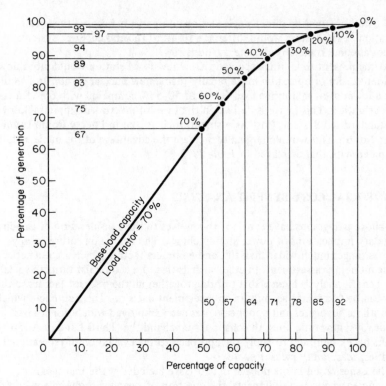

Figure 11.2 Annual electricity generation from UK power plant.

At the present time, there are insufficient nuclear stations to meet all the UK base-load requirements. They produce only 13% of total UK electricity, which will rise to 20% when stations currently under construction are completed and operational, at which time they will constitute 14% of the available generating capacity. *Figure 11.2,* also taken from Hunt and Betteridge[5], illustrates the relationship between installed capacity and the supplied percentage of total electrical output. Stations operating down to 40% load factor will comprise 70% of the total capacity and generate nearly 90% of total electrical output. As Hunt and Betteridge point out, the proportion of nuclear capacity could be increased by a factor of 4 if all base-load supply was from nuclear stations, and by a factor of about 7 before the break-even load factor indicated in *Figure 11.1* was reached and minimum system generating cost achieved. It seems unlikely that the national system would move this far, since the generating boards will wish to preserve some diversity in fuelling for a variety of operational reasons. At the present time, as stated in the CEGB Corporate Plan[6], it would be economic to install new nuclear generating plant built to standardised design before it is required to meet increased demand for electricity.

Consideration of the systems aspects of electricity generation in the UK will give a different measure of the economic benefit for installing a new nuclear power station from the direct single-station comparison, because the future operation of the system would differ as a result of the choice. Such systems studies are undertaken by the energy industries.

At the present time, it is expected that a new fossil station would move down the merit order of operation more rapidly than a new nuclear station and would have a lower average lifetime load factor of 50–55% compared with 65% for the nuclear station. This factor is derived in the form of investment appraisal known as 'Standardised Systems Cost Assessment' such as that in Energy Commission Paper No. 6[3]. This would marginally increase the advantage of the nuclear station compared with that displayed in *Table 11.1*.

11.3   SOCIAL COST–BENEFIT ANALYSIS

The above paragraphs have confined themselves to the consideration of the direct monetary implications of power station choice. The debate on future energy options ranges much wider than this and embraces factors which are not reflected in the monetary assessment. In cases such as this, the economist turns to social cost–benefit analysis in an effort to bring together all the relevant factors so that they can be taken into account in an investment decision. This approach, which has both its advocates and opponents, has been employed widely for studies on motorways, new tube lines, the channel tunnel and the 'Third London Airport'. Others have attempted to apply the approach to the evaluation of recreational facilities, road safety measures, etc.

The issues raised in the nuclear debate have included waste management, environmental and health impacts, the question of weapons proliferation and the related aspects of terrorism and civil liberties, and research and development costs.

I should say at the outset that I do not believe that meaningful monetary values can be attached to the environmental and proliferation aspects. However, I do believe that the social costs attached to pursuing the nuclear option are either smaller than, or not significantly different from, the costs of pursuing the fossil option alone, and, in any event, are small in relation to the overall economic benefit to be derived from the future use of nuclear power.

11.3.1   Radioactive Wastes

The spent fuel from nuclear reactors is highly radioactive, and United Kingdom policy is to reprocess this fuel to extract the re-usable uranium and plutonium and concentrate the fission products and other highly active wastes ready for conversion into a form suitable for ultimate disposal.

At the end of its useful life, a reactor or nuclear fuel plant will need to be dismantled and disposed of. The extent and rate of dismantling will be dependent on the value of the site and the costs at the time when the decision is to be taken. Inevitably, however, the costs of dismantling the nuclear plant will be higher than those associated with dismantling an equivalent fossil plant, and it is not unreasonable to include an allowance for this in an economic assessment.

Studies on the processes for disposal of radioactive wastes and for the decommissioning of reactors and plant have been conducted in many countries for a

148

considerable period and technical experts are confident that both can be accomplished without risk to the population or labour force*. If this is so, and I see no reason to doubt it, then the monetary cost of waste disposal and decommissioning of nuclear facilities can be built into the economic assessment, and this has been done in the figures presented in *Table 11.1*.

On the same basis, the costs of dealing with the wastes arising from coal-fired power stations and the costs of decommissioning fossil-fuelled power stations should be included. These costs are small in relation to the total system costs and make no significant difference. Environmental impacts arising from waste disposal and decommissioning are covered in section 11.3.2.

11.3.2 Environmental Impacts and Population Risks

Apart from visual intrusion, which would be broadly similar for either coal-fired or nuclear-powered stations at a given site, the environmental aspect of nuclear stations giving rise to expressions of concern is limited to the possible effects of low-level radiation releases or larger accidental releases on the workers in the industry and the general population. As everyone at this conference will know, the industry has an extremely good safety record. Reactors and fuel plant are designed to keep radiation emission levels down well below those considered acceptable by the appropriate international agencies.

Consideration is also given to hypothetical failure modes and their implications as part of the design process and a variety of means are adopted (design, warning, cut-out and containment) to ensure that the risks of any significant radioactive release are reduced to vanishingly small levels. The whole process is governed not by the designers and operators, whose responsibility for safety nevertheless remains, but independent inspectorates, without whose approval reactors and plant could not be operated.

Nevertheless, small quantities of radioactivity are released into the atmosphere and the risk of larger releases through accident can never be completely ruled out. Numerous assessments of the implications of these releases and risks for the workforce and the population at large have been made and several studies have attempted to set these in perspective against the risks attached to alternative means of generating electricity. The most recent studies have been those by Inhaber[7] and the UK Health and Safety Executive[8]. The latter confines itself to occupational deaths and omits occupational illness or impacts on the general population. Its results are presented in *Table 11.2*.

Inhaber[7] has attempted to take into account both health effects and general population exposures and extended his studies to include some so-called 'renewable' energy sources. *Figure 11.3* gives the comparison for fossil to nuclear plant derived from Inhaber and the HSE report, employing statistics appropriate to the United Kindom, insofar as they are available[9]. *Figure 11.4* presents Inhaber's data for the 'renewable' sources for comparison. The principal health effects in the case of renewables arise from the production of materials and the use of fuels

---

*See talk by Dr L. E. J. Roberts to BNES on 9 November 1978, entitled 'Radioactive waste policy and perspectives'.

TABLE 11.2 ESTIMATED NUMBER[8] OF DEATHS DUE TO ACCIDENTS PER
GIGAWATT YEAR (GW yr) OF ELECTRICAL ENERGY SENT OUT[a]

Primary energy source	Operation	Deaths/GW yr sent out (deaths caused by accidents)
Coal[b]	Extraction	1.4[c]
	Transport	0.2[d]
	Generation	0.2[e]
	Total	1.8
Oil and gas	Extraction	0.3[f]
	Transport	insignificant[g]
	Generation	none reported[e]
	Total	0.3
Nuclear	Extraction (USA)	0.1[h]
	Transport	insignificant
	Generation and reprocessing	0.15[i]
	Total	0.25

[a]Based on average electricity supplied from stations for years 1972–74 reproduced in Table 71, *Digest of United Kingdom Energy Statistics 1975* which shows relative outputs for nuclear, oil-fired and other steam-raising plants (of which the vast majority are coal-fired).

[b]Based on underground mining figures only. Open-cast fatalities are not included. (They would have an insignificant effect on the final figures.)

[c]Based on an assumed figure of 26 deaths in mining coal for power stations.

[d]Based on estimated number of deaths attributable to the movement of coal by rail not including accidents to the public. Figures for deaths due to the movement of coal by road considered insignificant.

[e]Based on number of deaths to CEGB employees 1970–77. Information supplied by CEGB.

[f]Based on figures for fatal injuries in exploration and production in the United States published in the *46th Annual Review of Fatal Injuries* report to the American Petroleum Institute. It is appreciated that the bulk of the fuel oil used in UK power stations is from the Middle East where accident performance may differ sharply from that of the USA. Figures for the Middle East are not available and we have therefore used the American figures as the only available indicator. Making allowances for the lower fuel-oil consumption of the UK, we have arrived at a figure of approximately two deaths per annum in overseas oil fields attributable to the production of oil for UK. To these we have added the figures for deaths in the North Sea oil extraction industry published by the Department of Energy. This gives an approximate figure of 12 deaths per annum.

[g]We have been unable to obtain figures for deaths due to the shipping of oil from overseas oil-producing countries.

[h]Based on figures from United States uranium mining reproduced in USAEC Wash 1224 *Comparative Risk-Cost-Benefit study of alternative sources of electrical energy.* It is appreciated that the USA is not the primary supplier of uranium for UK power stations. Again we have used these figures as they are the only ones readily available.

[i]Based on information supplied to CEGB (see f above) and by British Nuclear Fuels Limited for accidental deaths of employees 1970–77. None of these deaths was due to radiation effects.

required for the construction of the systems. Inhaber's own data include an allowance for back-up fossil-fuel plant to provide equivalent energy at times when the 'renewable' is not operative. This may not be appropriate in all cases and *Figure 11.4* presents the data both with and without the back-up.

150

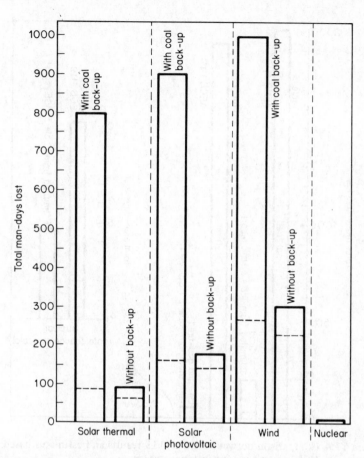

Figure 11.3 Comparison between risks of alternative energy systems with and without a coal-fired back-up system:——maximum estimate; - - -minimum estimate.

One may argue with Inhaber's procedure for bringing occupational deaths and general health effects onto a common scale, by treating deaths as equivalent to 6000 lost man days, but, regardless of this, it is evident that incorporation of environmental health and risk effects into a cost–benefit assessment would favour nuclear power over the alternative options. Even if an extremely high value was set to the human life*, it would not figure significantly in the monetary comparisons of *any* of the systems.

In addition to the population and health risk factors discussed above, a full cost–benefit assessment would include some allowance for more general environ-

---

*A practice which is followed in many social cost–benefit analyses, but one which I would not now consider appropriate.

151

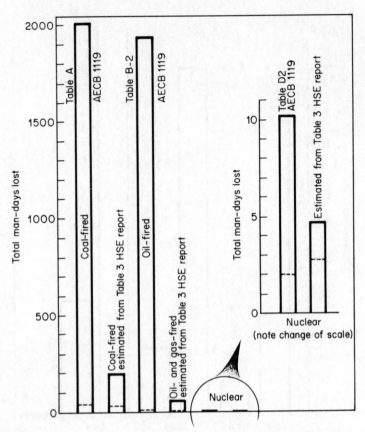

Figure 11.4 Comparison between AECB 1119 results and estimates based on the HSE report for conventional systems.

mental effects if these occurred. It is not obvious that there are any such effects associated with nuclear power, but acid oxide emissions from fossil-fuel combustion can contribute to corrosion and crop damage. Additionally, there may be effects associated with the emission of smoke or grit, mining subsidence and the movement and storage of large quantities of fuel.

Thermal pollution arising from the waste heat produced by either fossil or nuclear systems would be broadly equivalent and the effects would be negligible if the stations were located at coastal sites with direct sea-water cooling. It does not seem unreasonable to omit this factor from a comparative economic study. It should be mentioned that there can be some benefits associated with gaseous emissions and the availability of warmed cooling waters[19].

11.3.3 Proliferation, Terrorism and Civil Liberties

The effects of pollution are, in principle, quantifiable, though their precise impacts are not always fully understood and are site-dependent. The externalities covered

152

in this section are not susceptible to quantification in the same way. It is important to remember two factors. First, it is only the incremental effect associated with the addition of a nuclear reactor or reactors which should be considered and, secondly, that the consequences of adopting economically sub-optimal policies should be taken into account.

It is clear that the incremental effect on the spread of nuclear weapons of a decision by the UK to build one or more thermal nuclear power stations is negligible and for this reason the argument has only been raised in connection with the introduction of fast reactors or (elsewhere) the separation of plutonium for recycling in thermal reactors. Since any nation with reasonable technical competence could develop a nuclear weapons capability using much cheaper and more readily concealed routes than the adoption of a major civil nuclear power programme, it is difficult to see how it can be argued that the introduction of plutonium-fuelled reactors in the United Kingdom, already a nuclear weapons state, can materially add to the proliferation of weapons[20].

The existence of terrorists is no more related to nuclear power than is the existence of thieves to houses. Terrorism only has relevance in the social cost–benefit equation if nuclear power affords a means for terrorists to scale up their blackmail. I do not believe this to be the case. Nuclear installations are well protected against terrorist attack, so that there are far more attractive targets for terrorists which involve considerably less risk to themselves. If the present phase of world terrorism persists, the costs of protecting nuclear installations will appear in the overheads of the industry. These additional costs will, however, remain negligibly small.

The fear that precautions taken to protect installations against terrorists will infringe on the liberty of the individual citizen seems to me to be misplaced. Like proliferation, it is not relevant, nor is it argued, in relation to a choice of AGR versus a coal-fired station, but it is argued in relation to future programmes of fast reactors. Given much larger programmes of nuclear power, the numbers of protected sites would not be vastly more than those already existing for defence and other purposes. The existing sites do not appear to give grounds for concern and the marginal effect of the additional sites would be negligible. The principal concern expressed over existing sites appears to relate to their occupation of land, but this is a feature common to both nuclear and fossil stations. If terrorism were to become rife, then any additional precautions felt to be necessary would not be dependent upon the existence or non-existence of a sizable nuclear power programme.

For these reasons, I would argue that the whole proliferation, terrorism, civil liberties question is not only irrelevant to the decision to invest in an AGR or a coal-fired station but is also not a matter that would figure prominently in a full social cost–benefit study on the choice of direction for future UK energy strategy, even if it could be quantified.

11.3.4 R & D Costs

The issue of R & D costs is a complex one. Government expenditure on nuclear power R & D has been and continues to be considerable, though it is small in relation to the benefits likely to be derived from the introduction of significant quantities of nuclear power to the UK generating system. Each 2 GW of AGR on-line produces a net annual saving comparable to the total annual UK expenditure on reactor and fuel cycle development. It has not been a practice in the United Kingdom or elsewhere to include the costs of government-sponsored research and development on nuclear power in decisions of future power station investment, except to a very limited extent, nor is it appropriate to take past costs into account in investment decisions.

Most industrial companies charge R & D as an overhead on current activity and do not attribute its cost to individual end products in appraising forward commercial policy. It is not inappropriate that money expended to establish options for a future national energy strategy should be treated similarly and the costs of nuclear and non-conventional R & D treated as an overhead on total national energy expenditures. Some of the options into which research is undertaken may never be taken up, either because they are uneconomic or because circumstances change in some way; this is likely to happen to some, at least, of the 'renewables', and has happened to the steam-generating heavy-water reactor.

BNFL R & D costs are included in *Table 11.1* in the fuel costs. Energy Commission Paper No. 6 suggests that R & D and other launching costs associated with a new nuclear programme would, if spread over likely future programmes of reactor installation of the sort envisaged in the Working Document on Energy Policy, be equivalent to considerably less than 5% on the construction cost on each station. Inclusion of this sum would not significantly alter the comparisons given in *Table 11.1*.

If future R & D costs are to be included, then it would not be inappropriate to include future allowances for capital write-off, regional subsidies and expenditure on stockpiling for the fossil option.

11.3.5 Other Factors

In considering some of the factors that might be brought into a social cost–benefit analysis of the nuclear versus fossil-fuel option, we have introduced a number of wide-ranging issues that are more relevant to future overall energy strategy than to the choice of an increment to a generating system. Most of these have related specifically to nuclear energy. There are some equivalent issues, however, that relate to the choice of fossil systems or to the possible consequences of failing to develop the nuclear option.

The former is exemplified by the question of long-term accumulation of carbon dioxide in the atmosphere and its potential effects on world climate. That such a build-up has been occurring seems unquestionable (*Figure 11.5*). The significance and impact of such a build-up are difficult to predict or to attach any meaningful value to, either positive or negative[10].

The widely discussed studies of future world energy demand[11,12] express the view that the world will need to exploit vigorously all the technologies available to it, and energy conservation, if it is not to find itself in a period when economic growth is constrained by energy availability. The consequences of an energy 'shortfall' would appear as rising energy prices and depressed growth. If the shortages were sufficiently severe, it is not inconceivable that they could result in considerable international tension as countries vied with each other to preserve their own standards of living. If, on the other hand, nuclear power is

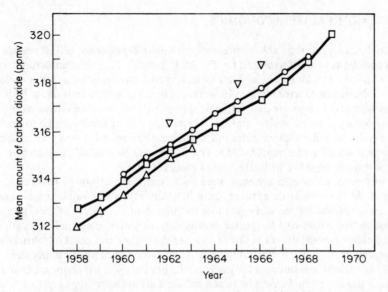

Figure 11.5 Increase in carbon dioxide measured at four widely separated locations (after Environmental Data Service): ○ Swedish flights; □ Mauna Loa; △ Antarctic; ▽ Barrow, Alaska.

exploited to the full, then it will help to reduce pressure on other energy sources and stabilise energy prices generally. If these scenarios are realistic and the path the world follows is dependent on the choices facing it, then the difference between the two scenarios would represent a benefit attributable to the development and exploitation of the nuclear option.

The benefits may not be immediately apparent after the course has been pursued, those of Dutch gas or our own North Sea oil are not immediately obvious to the man in the street.

11.3.6 Applicability of Social Cost–Benefit

From the above sections 11.3.1 to 11.3.5 it will be apparent that some of the indirect costs can be quantified in monetary terms and taken into account in specific investment decisions. Environmental and risk effects can be quantified in principle, though in practice this is a difficult task, but their conversion to monetary equivalents is controversial. Nevertheless, on the basis of the available evidence, they would appear to operate marginally in favour of nuclear power.

The issues covered in sections 11.3.3 and 11.3.5 are not relevant to specific investment decisions, but may be relevant to decisions on longer-term strategy. Neither section deals with matters that are readily quantifiable. In the author's view, the aspects covered in 11.3.3 are not significant in relation to the economic benefits associated with the nuclear option. The long-term implications for economic growth seem to be the most significant externality by far, and probably represent the greatest incentive to pursue the nuclear option vigorously.

155

11.4 FAST REACTOR ECONOMICS

It would be helpful to be able to present a complete break-down of fast reactor costs analogous to that presented for the AGR in *Table 11.1*. It is unfortunately not yet possible to do this in advance of having cost estimates for a fully developed series-ordered fast reactor station. However, experience with constructing PFR suggests that, although the first full-scale station will not be competitive with contemporary thermal nuclear stations, the margin will be small enough to give considerable promise that an economic design can be evolved through accumulating experience of actual construction. The figures will be available for the fast reactor inquiry together with site-related costs for CDFR.

Fast reactor economics are even more complicated than those of thermal reactors. At the present time, there seems little doubt that the liquid-metal-cooled fast reactor power station will cost more to build than an equivalent thermal reactor station because of its greater complexity. However, their fuelling cost per unit of electrical output will be lower, despite the higher unit costs of fabricating and reprocessing plutonium-bearing fuels. They avoid the costs of buying and enriching natural uranium and they achieve higher burn-up, which means that a smaller quantity of fuel has to be processed for a given electricity output.

Table 11.3 taken from Hunt and Farmer[13] illustrates a possible break-down of thermal and fast reactor generating costs.

TABLE 11.3 ILLUSTRATIVE BREAKDOWN OF THERMAL AND FAST REACTOR GENERATING COSTS assuming commissioning date 1998

		Thermal (%)	Fast (%)
Construction costs		55	67
Fuel cycle costs			
Uranium		13	–
Enrichment		7	–
Fuel fabrication and reprocessing		15	22
	Subtotal	(35)	(22)
Other operating costs		10	11
	Total	100	100

It seems likely that the early fast reactors will have capital costs which exceed the savings to be achieved on the fuel cycle, but with further development, manufacturing and operating experience the capital costs will be brought down to a level at which a fast reactor becomes the preferred option economically.

The particular advantage of the fast reactor rests in its ability to 'burn' 60% of the total uranium compared with only approximately 1% which can be burnt in thermal reactors, even allowing recycle of the plutonium.

The complicated trade-offs between aspects of reactor and fuel design and fuel cycle operation are described in the paper by Hunt and Farmer who stress the evolutionary nature of fast reactor development[13].

Just as in the case of a system made up of a mixture of thermal reactors and fossil-fuel stations, the fast reactor/thermal reactor/fossil-fuelled station mix for minimum operational costs will be determined by the relative capital and fuel cycle costs.

The rate at which fast reactors can be introduced into the UK system will be dependent on the stocks of plutonium that have been built up by prior operation of thermal reactors, by the breeding gain in the early fast reactor systems and by the time cycles practised in fuel reprocessing. Initially, the inventory will be the more important since the reactors will be fuelled from existing stocks, and designs with the smallest inventories will allow for more fast reactors to be fuelled from those stocks. This has an immediate impact on uranium requirement because each GW of fast reactor will reduce uranium requirements by some 4000 tonnes over its lifetime compared with the equivalent thermal reactor capacity.

The attainment of small inventories depends not only on reactor design but also on the fuel cycle and it is important to minimise the hold-up in the recycling of spent fuel to a minimum.

Once initial stocks of plutonium have been consumed, the doubling time becomes the dominant factor controlling the rate at which fast reactors can penetrate the UK system. The percentage annual growth rate of fast reactors will equal the inverse linear doubling time. Provided this is equal to or greater than the required rate of growth in the total nuclear component, thermal reactors could be phased out completely. It will take several decades before the UK electricity system could become independent of thermal reactors for base-load operation, after which supplies of depleted uranium will be adequate for meeting the country's electricity requirements for several centuries.

Authoritative studies on uranium availability give grounds for believing[14] that lower-grade resources will have to be used increasingly from some time around the turn of the century. This, allied with likely increases in the real costs of fossil fuels and their consequent effects on the separative work component of nuclear fuel costs, will lead to increases in the operational costs of thermal reactors. Since the fuel costs currently only amount to some 15% of the total costs of nuclear generation (*Table 11.1*) significant increases could be tolerated without upsetting the nuclear/fossil trade-off. In a hypothetical free and rapidly responding market, the uranium price would move up to the point where the prospective costs of operating thermal and fast reactors were equal at the margin and would then stabilise. In practice, the situation is complicated because of the lifetime requirements of fuel for existing reactors and by the structure of the market, which operates through long-term forward contracts, with a comparatively small proportion of fuel purchased in the spot market. Nevertheless, the successful development and introduction of fast reactors would be expected to stabilise uranium prices and the costs of electricity generation and also contribute to the stabilisation of the costs of other fuels. The full benefit attached to the development of the fast reactor option is therefore extremely difficult to measure since it is related to the whole question of world growth as indicated earlier.

157

11.5 ACCOUNTANCY

I have concentrated in this paper on the economic assessment of future investment options rather than on accountancy statements related to the current costs of generation. Much confusion arises from a failure to understand the accountancy conventions, the differences in conventions between countries and the manner in which changes have been introduced over the years.

Table 11.4 presents the CEGB figures for generating costs based on historic costs expressed in money appropriate to the year of generation[15]. The comparison is limited to systems built during the preceding 12 years but differences in the availability of individual stations can affect the comparisons.

TABLE 11.4 CEGB GENERATING COST DATA

| | Generating cost (p/kWh) of stations constructed during previous 12 years (in current money terms) | | |
	Oil	Coal	Nuclear
1971/2	0.39	0.43	0.43
1972/3	0.40	0.49	0.48
1973/4	0.55	0.53	0.53
1974/5	0.88	0.74	0.48
1975/6	1.09	0.97	0.67
1976/7	1.27	1.07	0.69
1977/8 (provisional)	1.42	1.23	0.76

Transmission and distribution costs more than double the cost of electricity to final consumers.

Table 11.5[16] gives a break-down of costs, but even here changes in presentational form have taken place. The figures for 1974/5 were confined to current costs, whereas those for later years included an allowance for commitments to be met in future years. Neither set of figures are a guide to future investment.

TABLE 11.5 BREAK-DOWN OF COSTS (p/kWh)

| | 1974/5 | | | 1975/6 | | | 1976/7 | | |
	Nuclear	Coal	Oil	Nuclear	Coal	Oil	Nuclear	Coal	Oil
Fuel costs	0.13	0.55	0.71	0.25	0.75	0.87	0.34	0.86	1.05
Other operating costs	0.09	0.07	0.05	0.14	0.08	0.07	0.11	0.09	0.08
Capital charges	0.26	0.12	0.12	0.28	0.14	0.15	0.24	0.12	0.14
	0.48	0.74	0.88	0.67	0.97	1.09	0.69	1.07	1.27

I do not propose to go into detail on the accounting practices of the generating boards, which are adequately covered by Hunt and Betteridge[5]. The historic cost comparisons are based on standard accounting conventions used generally throughout industry and, with no inflation, this approach would give a reasonable approximation to future generating costs. The adoption of Current or Replacement Cost Accounting has been discussed and is intended to ensure adequate accumulation of funds for replacement of capital assets and stocks. The generating boards have recently adopted a procedure which involves a 40% increase in depreciation provisions which would narrow but not eliminate the gap between historic costs of nuclear and fossil stations presented in *Tables 11.4* and *11.5*.

It may be helpful to illustrate some of the types of erroneous thinking that appear in the press and literature as a result of failure to appreciate the significance of accountancy figures.

The maintenance error that led to the shut-down of Hunterston 'B' reactor has led to an apparent increase in the costs of generating nuclear electricity, as expressed in the SSEB accounts, because the capital charges for a new power station are conventionally included after it has been first commissioned but not before. The Scottish figures therefore show a temporary aberration in the nuclear generating costs associated with the debiting of full capital charges spread across a reduced output. This has nevertheless been used erroneously as a basis for questioning the competitiveness of nuclear investment in Scotland.

A letter to the *Guardian* in August 1978 sought to question the AGR fuel cost estimates contained in Energy Commission Paper No. 6 by comparing them with dissimilar data drawn from the CEGB statistical year book. The former relate to AGR fuel which will have a higher burn-up, whereas the latter relate largely to Magnox fuel. This mixing and comparing of unlike figures, or comparison of figures taken for different years without due reference to the effects of general inflation (often selectively), can lead to totally erroneous statements. The US Ryan Report[17] presents many classic examples of cavalier treatment of cost data to which attention is drawn in the minority reports.

11.6 CONCLUSIONS

I have sought to give a fair picture of the economic case for nuclear power. The costs presented are based on the most recent and detailed assessment available in the UK, but inevitably they include judgements, for example on the future trends in fuel prices. For the reasons presented in the text, there are inevitably uncertainties and these will lead to some differences of view on precise relativity of fossil- and nuclear-fuelled power stations. Nevertheless, there is general consensus within all the fuel industries that nuclear power offers the cheaper route to meeting the base-load demand for electricity. There is also general agreement that the UK will need to base its forward electricity generation programme on a mixture of coal and nuclear stations and that neither alone would suffice. There would be more disagreement on the precise break-even load factor between coal and nuclear and, hence, the optimum minimum cost mix. Nevertheless, the divergence between the industries does not appear to be all that great.

The economics of fast reactors are complicated and at this stage it would be premature to present a detailed economic case. There is no doubt that the proposed commercial demonstration fast reactor will produce electricity at higher costs than an advanced gas reactor or pressurised water reactor but probably at lower costs than contemporary fossil stations. The time when fast reactors become the economically preferred option against thermal nuclear stations will depend on the rate at which the capital costs can be brought down, and the prospective costs of uranium fuel for thermal reactors over their lifetime. Extreme scenarios with political constraints on uranium supply could bring such a situation about in the 1990s. On the other hand, fresh discoveries of large low-cost uranium deposits could defer the date at which the option became attractive. It is because of these uncertainties that the sensible course for the UK appears to be construction of a commercial-scale fast reactor to gain operating experience on the reactor and its fuel cycles. Decisions on whether and when to go ahead with the introduction of a series of fast reactors could then be made in the light of circumstances as they develop.

The text has examined some of the topics raised by the opponents of nuclear power and fast reactor development and sought to set these in perspective. It is not evident that any of the factors could be regarded as adding significantly to the costs of nuclear power in a full social cost–benefit assessment. This is fortunate since externalities are notoriously difficult to quantify and some of those raised in the nuclear debate virtually impossible. The biggest unknown is, perhaps, the benefit to be attached to the development and application of nuclear power, in terms of the effects of lower energy costs (in relative terms) on general UK and world economic development. This is an aspect that deserves more study than it has received to date.

REFERENCES

1. NNC Report, *The Choice of Thermal Reactor Systems*, 1977
2. CEGB Announcement (see *Financial Times*, 4 November 1978)
3. Energy Commission Paper No. 6 (1978)
4. Cmnd 7131, HMSO, London (1977)
5. Hunt, H. and Betteridge, G. E., 'The Economics of Nuclear Power', *Atom, December 1978*
6. CEGB Corporate Plan, 1978, p. 31
7. Inhaber, H., 'Risks of Energy Production', *Atomic Energy Control Board (Canada) Report AECB 1119*, March 1978
   Jones, P. M. S., *Atom*, August 1978, p. 223
8. *The Hazards of Conventional Sources of Energy*, Health and Safety Executive, HMSO, London (1978)
9. Williams, D. E., unpublished studies
10. Smith, I., 'Carbon Dioxide and the Greenhouse effect', *IEA Coal Research*, ICTIS/ER 01, April 1978
11. *World Energy Resources 1985–2020*, World Energy Conference, IPC Press (1978)
12. *Energy, Global Prospects 1985–2000*, Workshop on Alternative Energy Strategies
13. Farmer, A. A., and Hunt, H., 'The Important Factors in Fast Reactor Economics', *Nuclear Engineering International*, July 1978

14. *Uranium Resources, Production and Demand*, OECD/IAEA, January 1978
    Nuclear Fuel Cycle Requirements, OECD/IAEA, February 1978
15. Replies to Parliamentary Questions
16. *Electrical Times*, 2–9 June 1978
17. Jones, P. M. S. and Sargeant, J. 'Nuclear Power Costs', *Atom*, October 1978, p. 280
18. *Nuclear Power Costs*, US Congressional Report, April 1978
19. Jones, P. M. S., Taylor, K. and Storey, D. J., 'A Technical and Economic Assessment of Air Pollution in the UK', *Programmes Analysis Unit Report* PAU M20, HMSO, London (1972)
20. Sir J. Hill, *Royal Institution Forum Papers on Nuclear Power and the Future*, p. 217 et seq
    Jones, P. M. S. 'Nuclear Energy Prospects', *Atom*, February 1978

12

Critique of the Economic Case for the Fast Reactor

Colin Sweet

12.1 INTRODUCTION

The debates in the Power Steering Committee at Harwell during the formative stage of the early reactor programme were marked by a good deal of confusion as to what were the objectives in reactor design[1]. The confusion arose because there were several objectives in the minds of those responsible for policy and, as is often the case, they were not compatible in practice. This confusion has not only persisted, but today is far more obvious than it was 30 years ago. Nowhere is this more evident than in the area of nuclear power economics.

In the minds of the scientists at Harwell, the building of the fast reactors was the main goal of the nuclear programme and there were few doubts about that. It was a view shared with the Americans. The first reactor in the world to generate electricity was to prove to be a fast reactor—and so was the first reactor to be shut down because of a serious accident. Both were in the USA.

The scientists and engineers at Harwell and Risley, however, were not euphoric about the economic benefits of fast reactors. They were very cautious in their estimates. They 'never held out the promises of cheap power from fast reactors'[2]. In their view, the strategic advantage lay in overcoming the constraint that the scarcity of uranium placed on nuclear power. 'They realised that, in the long term, calculations of the fast reactor's financial advantages compared with those of thermal reactors would depend in part on the scarcity and price of natural uranium and of the pure fissile material ^{235}U and plutonium'[3].

It is also perhaps worth emphasising that the objective of the exercise at first being considered was not to produce a fast reactor as such, but to produce a breeding reactor. These two are not necessarily the same.

Breeding in fast reactors has, in fact, turned out to be more difficult and costly than might have been expected. But breeding or near-breeding can be achieved without fast neutrons. One of the merits of the Magnox design introduced in the first UK programme is that it has a high plutonium yield. The Candu reactor, chosen by the Canadians, which has an even better neutron economy, could

163

achieve breeding using thorium, which is at least as abundant in the Earth's surface as uranium. By contrast, the light-water reactor (LWR) and the advanced gas-cooled reactor (AGR) are heavy users of uranium relative to the amount of plutonium that they yield. The choice, therefore, in Britain to follow most other countries and to opt for enriched uranium reactors (AGRs) was to accelerate the drain on uranium supplies and made it imperative to go to the fast breeder reactor as soon as possible. This assumed that uranium at an economic price could become scarce before the end of the century, a view which is being currently questioned as the supply/demand pattern begins to suggest that reserves might well last into the next century.

In the second stage of reactor development, when the second generation of thermal reactors (AGRs) and the Prototype Fast Reactor (PFR) were both approved and commenced construction, the apparent necessity of low-cost performance was added to that of reactor feasibility. In the first stage, the objective had been to design reactors that would use fuel efficiently and yield a good power output for the generation of electricity. The first commercial reactor was not cost competitive and it was a legitimate objective that the Magnox series should demonstrate their competitiveness before more nuclear stations were ordered.

The acceleration of the commercial programme, before this had been achieved, precipitated an economic case for the AGRs that had to be made by *force majeure*. In this new climate, the same demand was made for the fast breeder although it was nowhere near the stage of commercial development. Thus, in the space of a decade, the designers and engineers had moved from a situation in which the practical requirement for which reactors were being innovated had been to produce plutonium for military purposes, and for which the only consideration was one of technical feasibility, into an entirely new situation in which cost competitiveness was placed alongside technical feasibility.

Hitherto, the cost had remained a largely unknown factor, because the technology (especially that of the fast breeder) was at an R & D stage, and economic science has no reliable method for measuring R & D outputs. The insistence of the policy makers in the early 1960s upon economic feasibility changed the path of nuclear development. By this insistence, engineering feasibility and economic feasibility became locked together. The result was, of course, disastrous for the prospect of any clear thinking about the future.

The place that nuclear power with its complex technology and the considerable number of technical unknowns and production variables ought to have found within the sphere of fuel and power economics (which in any case was dominated by the rather purist marginal cost analysts) would in any case have been difficult to establish, except over a considerable period of time. The failure of the political decision makers of the 1960s to insist on economically rigorous methods for testing nuclear power was compounded by the fact that neither the nuclear power industry nor the electricity supply industry possessed the personnel, the right approach or the decision-making structure that was capable of handling such problems. For its own reasons, the Government abdicated its responsibility and allowed them to believe that they had all the requisites for realistic decision making.

The result was a rapid escalation of the economic claims not only for thermal nuclear stations but for the fast breeder as well, notwithstanding its early stage of development. These claims were accompanied by statistics which were misleading —not intentionally so, but because the advocates were themselves misled. The technical uncertainties which would have made the best economic assessments perhaps little more than intelligent guesses remained as hidden rocks, submerged from view for the purposes of political necessity. For the time being, at least, economics as a relevant social science in this part of the public sector was relegated.

The results which became visible by the mid 1960s were a series of projections based on an oversimplified and wish-fulfilling manipulation of statistics. When the result of the Dounreay Experimental Reactor became available, a large-scale and rapid expansion of the fast breeder was urged. Mathews and Frame, director and chief engineer, respectively, of the Dounreay Reactor, were quite clear about the economic gains that would be realised. They were fully supported in an article written at the same time by Franklin and Kehoe of the Atomic Energy Authority.

Mathews and Frame in their article entitled *Fast Reactors On Line by '71* stated that capital cost would be no greater for fast breeders than for thermal reactors. They[4] put the cost as low as £60/kW and possibly falling to £50/kW! Franklin and Kehoe had stated that installed electrical capacity by the end of the century might be 120 GW (compared with 36 GW for 1966)[5]. Mathews and Frame were specific about nuclear power's role. 'Typically, in a nuclear power programme rising to 50 000 MWe by 1985 and to 100 000 MWe by the end of the century, it is conceivable that the UK could include 10 000–20 000 MWe and over 75 000 MWe, respectively, of fast breeder reactors'[6].

Such was the misplaced euphoria of that time, that engineers with such practical experience could believe that it was possible on engineering grounds to build an enormous fast breeder programme in so short a time. They should have been reminded of what the engineers in charge of the very earliest fast breeder work wrote: '. . . at first sight the reactor appears unrealistic. On close examination it appears fantastic. It might be well argued that it could never become a serious engineering proposition'. Neither at that time, nor since, has any of the bodies responsible for the development of the fast breeder shown that it could be a serious economic proposition.

12.2 CRITIQUE OF THE COST STEREOTYPES

Because of the peculiar history of nuclear power economics, the background to which I have tried briefly to outline, we have a very unsatisfactory situation today with what purports to be estimates of fast breeder economics. Although virtually no reliable economic data exist on the fast breeder, we nevertheless have, emerging in the literature, a treatment which, in the absence of the necessary information, is loosely based on the structure of fast breeder costs. From this treatment, conclusions are drawn about the overall cost competitiveness of the reactor. While discussion of the structure of fast breeder costs is perfectly legitimate, in the absence of any actual costs it can only be treated as a very tentative exercise. To place the structure before the data on which it ought to be based is equivalent to espousing a theory before there is any evidence to support it.

165

That this can happen is the result of what I call the nuclear cost stereotype and which is a product of the peculiar history of nuclear economics. In general, the cost stereotype asserts that the high capital cost of nuclear power compared with the lower capital costs of fossil-fuel systems is more than counterbalanced by the low fuel costs in nuclear power and the relatively high fuel costs in fossil-fuel systems. While the stereotype is not devoid of some historical truth, it may be wrong in detail, and it may show a very large margin of error which expands over a period of years. It is static, often manipulative and, instead of being the product of serious research, is far too often advanced in support of a conclusion which is not based on any worthwhile evidence.

A recent example of the cost stereotype as applied to the fast breeder is that given in *Table 12.1*, and advanced by Hunt and Farmer of the Atomic Energy Authority[7]. These figures relate to stations to be commissioned in 1998 and are described as 'a typical breakdown of thermal and fast reactor generating costs'[8].

TABLE 12.1 RECENT EXAMPLE OF THE COST STEREOTYPE APPLIED TO THE FBR

	Thermal (%)	*Fast* (%)
Capital costs	55	67
Fuel costs	35	22
Other operating costs	10	11
	100	100

How this table is derived we are not told, and why it should be valid at the end of the century is even less apparent. But it is intended to be a genuine cost stereotype to which it is believed the industry will respond, and which will accommodate market prices with the right result never in doubt. The accompanying text to the above table argues: 'These figures illustrate the importance of capital costs. Because thermal reactors will have to bear prices for uranium and enrichment adequate to encourage expansion of supply, fast reactors could cost more than thermal reactors and still remain competitive'. At first we are told fast reactors may not be able to operate within the allowable margin, but with development 'construction costs should be brought within the required margin'. The interpretation of the stereotype is that the allowable margin for capital costs of the fast reactor is in fact determined in part, if not wholly, by the relative upward movement in fuel costs for the thermal reactor. Thus: 'Also the prices of uranium and separative work (enrichment) are expected to continue to increase as prices of competing fossil fuels go up. Fast reactors being immune to these prices will help to stabilise the cost of electricity'.

The juxtaposition of the hope that uranium prices will rise with the naive belief that fast reactor fuel costs will not, recalls equally incorrect but opposite statements that were common in 1974/5 to the effect that nuclear power was immune from the rising prices triggered off by the OPEC price rise of 1973/4 because the price of uranium was not affected, and that nuclear power was price insensitive on the fuel supply side. Too much play has been made, and continues to be made, with nuclear prices by writers anxious to place nuclear power in the most favourable light. Not many years ago they were held to be static, which was unrealistic. Now they are held to be exceptionally volatile, which is equally unrealistic.

The unsuitability of analysing the case for thermal or fast reactors by way of stereotypes arises, however, from something more than the difficulty of trying to make the cost data fit the structure. In general, we would agree that a comparison of fast and thermal reactor costs with fossil-fuel costs is a proper one to adopt. The objection to the cost stereotype method of doing this is that, in order for it to have any meaning as a guide to the future, it requires to have built into it a relative fuel price effect which it is assumed will hold for as long as the prediction is meant to hold. This assumption then acquires the force of necessity. Or, rather, it had to acquire the force of necessity once the prediction in the middle 1960s[9] that the capital cost of nuclear stations would fall to a level similar to those for coal was proved hopelessly wrong. Once the expectation of falling capital costs was contrasted with their sharp movement in the opposite direction, then the only prospect for nuclear being competitive was that fuel costs would remain low enough to offset the high capital costs. (Low, that is, relative to the fuel costs of fossil-fuelled power stations.) The logic of the stereotype (taken in isolation) is irrefutable. Fuel costs of nuclear stations must remain exceptionally low in order to counter the exceptionally high capital costs. Logic, however, has no necessary connection with reality.

The cost breakdown in *Table 12.2* for nuclear and fossil fuels is given by Hunt and Betteridge[10]. If we convert the figures for nuclear stations in the table into a percentage cost structure table, we get *Table 12.3*. If this table is compared with that of the stereotype of Hunt and Farmer (*Table 12.1*), it will be seen that, in the three years covered, the actual costs have moved away from the stereotype in a dramatic way. In 1976/7 fuel costs have risen to almost half the total costs and capital costs are down to a third. The trend continues into 1977/8, by which time the cost structure becomes the exact reverse of what the stereotype requires. Hunt and Farmer do not agree with Hunt and Betteridge!

These authors might argue in defence of their stereotype that the inversion that I have demonstrated arises because Magnox stations are not typical, the capital

TABLE 12.2 COST BREAKDOWN (in p/kWh) FOR NUCLEAR AND FOSSIL FUELS

	1974/5			1975/6			1976/7		
	Nuclear	Coal	Oil	Nuclear	Coal	Oil	Nuclear	Coal	Oil
Fuel costs	0.13	0.55	0.71	0.25	0.75	0.87	0.34	0.86	1.05
Other operating costs	0.09	0.07	0.05	0.14	0.08	0.07	0.11	0.09	0.08
Capital charges	0.26	0.12	0.12	0.28	0.14	0.15	0.24	0.12	0.14
	0.48	0.74	0.88	0.67	0.97	1.09	0.69	1.07	1.27

TABLE 12.3 PERCENTAGE COST STRUCTURE TABLE FOR NUCLEAR STATIONS
IN TABLE 12.2

	1974/5 (%)	1975/6 (%)	1976/7 (%)
Fuel costs	27	37.5	49.2
Other operating costs	18.7	20.8	15.9
Capital costs	54.1	41.8	34.7
	100	100	100

costs having been undervalued. If this is so, it might be timely after 17 years of Magnox operation (as the low-cost option in electricity generation) for the generating authorities to elaborate the point. If the real cost for Magnox generation is substantially divergent from the published costs, this has an important bearing on the evaluation of all nuclear costs for the UK.

Turning to fast reactors, it can be readily understood that if there was a strong element of necessity in arguing that the cost stereotype had to show low fuel costs in order to offset the high capital costs for thermal reactors, the same argument has to be applied even more strongly to the fast reactor. All sources agree that fast reactor capital costs will be substantially higher than thermal reactor capital costs. The only question is by what amount. The lowest figure published is 25%[11] and, as this was based on a study made in 1974, and carried out mainly on the slide rule, it has little relevance today. The present favoured guideline is that of the French Commercial Fast Breeder, Superphenix. Dr Walter Marshal has suggested, using the French case as a relevant analogy for the British CDFR, that 160% of thermal reactor costs would be a proper guide for the UK reactor. Latest estimates put the Superphenix at twice the cost of the equivalent thermal reactor[12]. If fuel cycle costs rise in real terms, which is likely, then the requirement of a relative diminution in fuel costs will not be met, with the result that total costs will rise by at least as much as capital costs rise. The stereotype does not so much deny this possibility as conceal it from view. Hunt and Farmer used the rising cost of uranium as their exogenous variable which reduces the relative competitiveness of thermal reactors. This, in turn, is derived from another exogenous variable—world-wide demand which is particularly difficult to estimate. They then attempt a trade off by arguing that fast reactor capital costs could rise provided the price effect of the uranium market on thermal reactors was the greater of the two.

As Superphenix is not due to be completed until 1983, the above figure can be regarded as a minimum rather than a maximum figure. The French authorities are acutely conscious of the mounting cost of Superphenix and are devising a variety of ways by which they hope to reduce the costs for further reactors of the same size[13].

If we then assume in the light of our present very limited knowledge of fast reactor costs that the capital cost will be twice that of thermal reactors of the same size, then it follows that the fuel and operating costs must be a fraction of the fuel costs of thermal reactors if the reactor is to be competitive. The gearing of these costs is interesting. If capital costs are 70% of total costs in the stereotype, and fuel and operating costs 30%, and if capital costs should rise by 1% of total costs, then fuel costs have to fall in real terms by a factor of 2.3, if total costs are to remain the same. That is to say, the higher the capital costs become, the greater the proportionate amount by which fuel costs must fall, if there is a total cost constraint.

Further, in order to deduce then that fast reactors are competitive, Hunt and Farmer have assumed, as the cost stereotypes normally do, that thermal nuclear is cheaper than fossil-fuelled stations. There is, however, no reason for making this assumption. If thermal nuclear costs rise above those of fossil-fuelled stations for the same load factor, whatever the cause, then there is no reason for preferring thermal stations. The real comparison therefore has to be made with fossil-fuelled stations. This has been done by the Department of Energy[14], but for some un-explained reason they assume a rise of 3% per annum in real cost terms for coal in the maximum case and a near-to-zero increase in the cost of nuclear fuels in the nuclear case. This contradicts what the AEA have been saying and what BNFL said at the Windscale Inquiry[15] to the effect that the world price of uranium would rise very sharply, and record a large overall upswing in the next two decades.

The implicit argument attached to the use of cost stereotypes has been that thermal nuclear stations are more economic because their fuel cycle costs are low. Economic, that is to say, relative to fossil-fuel reactors. Now it is argued that fast breeders will be preferable because the fuel costs of the thermal reactors are rising and will continue to rise by substantial amounts. If that is true, then thermal reactors will be more costly to build and operate than fossil-fuelled stations, and indeed there is evidence to substantiate this point[16]. That may be correct. But what is not supportable is to argue that low fuel costs are a boon to the thermal reactor or *vis à vis* the fossil-fuelled station, and then to argue that high fuel costs for thermal reactors are a boon to the fast reactor case.

Does the cost stereotype have any meaning? Only to confuse the issue. Low fuel cycle costs for nuclear power cannot be sustained as a characteristic feature once the stereotype is adapted to accommodate the case for the fast reactor. If fuel cycle costs are low for the thermal reactor, then there is no case for the fast reactor. On the other hand, if fuel cycle costs are up for the thermal reactor, thus giving licence to the fast reactor, there is no case for the thermal reactor, because fossil fuel stations would be preferred. Because it is being used to argue mutually contradictory propositions the stereotype is not useful.

Comparative cost statements, especially for large technological projects, have meaning only if taken over a reasonable part of their lifetime, which is one of the reasons why such statements are difficult to make. We also have to assume that real costs approximate to market prices and that capital scarcity will be resolved in conditions where opportunity costs can be stated with some degree of accuracy. Assuming that prices are determined in some kind of market, and not administra-tively, we would argue that, while it may be formally true in terms of the cost stereotype for fossil fuel costs and nuclear power costs to move in separate paths (not excluding opposite directions), it is a mistake to think that in practice this will happen. If there is a sharp rise in fossil fuel costs, then it is unreal to believe (as nuclear analysts did in 1974/5) that nuclear costs will be unaffected. The difference between different fuel costs will be visible at the margin, and in response to the rise in the marginal energy supply price, i.e. the world oil price. Similarly, the attempt to argue into the cost stereotype a significant difference in price movements for thermal and fast reactors is open to the same objection. While technical factors affect cost structures, costs are market-determined and not technologically determined. There will be only one market for nuclear fuels and it is a rising one. Given that the issues are marginal ones, it would seem logical for nuclear power analysts to explore the differences between plutonium recycle in thermal reactors compared with recycle in fast reactors. As Marshall has made clear, however, this is not a path that has been seriously pursued to date in the UK.

The inescapable conclusion of this rather formal treatment of the subject of the fast reactor cost is that, because of the high capital costs, the viability of the reactor in economic terms must depend on the fuel cycle. There is, as yet, no evidence to give credence to the argument that there will be any economies of scale in fast reactor construction. The official policy is to build one Commercial Demonstration Fast Reactor before any longer-term conclusions can be drawn. That policy also concedes that the first commercial reactor will be built at a loss. Whether this is acceptable is questionable, and will no doubt be thoroughly debated at the forthcoming inquiry. But all the evidence points to high costs as being inescapable. The argument, therefore, for the viability rests on economies in the fuel cycle as the countervailing factor against the high capital costs.

12.3 THE FAST REACTOR FUEL CYCLE

The principal difference between the operation and maintenance of thermal reactors and fast reactors are three-fold, as now discussed.

12.3.1 Reprocessing and Refabrication Costs

Because fast reactor fuel has a much higher burn up (100 000 MW days per tonne compared to 30 000 MW days per tonne for a light-water reactor), it has to be taken out of the reactor repeatedly to be reprocessed and refabricated, and then returned to the reactor. For economic reasons, the time out of the reactor has to be kept as short as possible, i.e. desirably, one year or less. Reprocessing thermal reactor spent fuel is still in the development stage. Nowhere in the (non-communist) world is there a large reprocessing plant for LWRs or AGRs with established experience. Reprocessing of fast reactor fuel is more difficult because of the high irradiation of the fuel in the reactor. While this could be reduced by allowing for lengthy cooling times, as with LWRs, the necessity of rapid recycling of the plutonium back into the reactor precludes this option. The resulting technology is not only complex and more hazardous but it is more expensive. The same is true for refabricating.

Reprocessing costs for LWRs have risen by a factor of 10–20 over a 15 year period. For fast reactors, the costs could be twice as high. Refabrication costs will be of a comparable order. Reported cost figures and estimates vary greatly.

The real cost of reprocessing oxide fuels at THORP will be substantially above the £250 000 per tonne quoted by BNFL at the Windscale Inquiry. The Parker Report states that the plant will run at a loss. Estimates for reprocessing fast reactor fuel in the UK are now as high as £500 000 per tonne and refabrication costs are even higher than this. Waste disposal costs must be added, plus decommissioning costs (which have been placed by BNFL at £50 million for the THORP plant).

The capital cost of reprocessing and waste disposal plants is high. BNFL claimed that THORP would cost £600 million. My own estimate was approximately twice that figure. The cost of the Gorleben plant for reprocessing and waste disposal in West Germany was placed at £3000 million.

While we do not know the costs therefore of the fast reactor fuel cycle, we know that they are going to be high, and it remains to be demonstrated whether they will be lower than the thermal reactor fuel cycle. This may depend on what happens to uranium costs. The expectation of the rise in uranium prices is based on supply and demand. Because demand for nuclear power has faltered badly and supply has increased, it is very likely that uranium and enrichment prices will flatten off much earlier than has been expected. At the same time, it is certain that reprocessing costs, which now comprise approximately one-third of fuel cycle costs in thermal reactors, will be a high cost in the fast reactor cycle together with refabrication. The back end of the fuel cycle costs will exceed the front end costs and it remains to be seen whether this will more than offset the cost of uranium in the thermal fuel cycle.

12.3.2 Out-of-Reactor Time and Economic Optimisation

The second feature of fast reactors is that the optimisation in the reactor management policy is very different from that in the thermal reactor. Fast reactor fuel cost per unit weight of fuel is much higher. It is generally agreed that the maximum burn-up to be aimed at is 10%, i.e. 100 000 MWe/tonne. Such heat outputs raise problems which in turn make the economic optimisation problematic. Plutonium metal is not considered capable of such performance. The search for alternatives has led to mixed plutonium/uranium oxide fuels and carbide fuels. Work on these advanced fuels continues and the results will affect the performance of the fast reactor and, in particular, its doubling time. Maximising the fuel rating through fuel design is, in certain respects, incompatible with shortening the doubling time. The design based on the small-diameter fuel pin has a high specific power and a smaller reactor inventory, but it achieves a relatively small breeding gain. It relies on frequent reprocessing to achieve a better doubling time. This means the out-of-reactor time has to be shortened and this raises serious difficulties. The larger-diameter fuel pin has a lower specific power and a higher breeding ratio, with a result that the reprocessing is less frequent and the doubling time is higher.

The incompatibility between a short doubling time and maximum heat output is not only a result of the time out of the reactor, but also the result of raising the plutonium enrichment in the fuel pins in order to raise the thermal rating which reduces the available neutrons for capture in the blanket. Similarly, the result of increasing the burn-up time by keeping the fuel in the reactor for longer periods

171

involves a trade-off between breeding and unit output costs. The relation between the factor and the simple doubling time (in years) is shown by the following formula

$$\text{doubling time} = \frac{10^6 M}{365PL\,(b-1)\,(1-f)}$$

where M is the mass of fissile material in the fuel cycle, P is the heat output in megawatts, L is the load factor, b is the breeding ratio and f is the fast fission bonus.

The raising of the value of P may reduce the value of b or f. The power output in the blanket may be reduced to a point where it is more costly per kWh than in a thermal reactor. The achievement of a high load factor is also important and most studies, it should be noted, assume a load factor of 75%, which is too high.

The overall result, therefore, of achieving the maximum heat output will be to reduce the cost per unit of power generated, but it also reduces the breeding gain and the doubling time. Or, to put it another way, the breeding gain is inversely related to the cost. A possible, but partial, solution is to improve the amount of breeding in the core. The decisions, therefore, on the parameters used in optimising the fast reactor depend on what are understood to be the desirable economic benefits.

12.3.3 The Doubling Time and the Logistics of the Fast Breeder Programme

Until a few years ago, the literature on the fast breeder reactor was almost universal in its emphasis of the benefits that breeding by the LMFBR would bring. Perhaps the apogee of the claims then being made was that by Chauncey Starr and Hafele to the effect that even if the price of uranium rose by a factor of 10 000 the fast breeder would be an economic power producer[17]. Today the emphasis has entirely changed, and nothing could illustrate more the confused perspective that prevails. In place of the short doubling time (7 to 15 years), we have in almost all studies from the nuclear industry and supporting agencies a long doubling time (25 to 50 or more years). Breeding has, in the case of Marshall's scenario, become more of an option than a necessity[18].

A number of reasons can be specified for this present emphasis.
(1) The earlier scenarios were based on the concept of the rapid replacement of thermal reactors, thus lifting the uranium supply constraint.
(2) The incompatibilities in the fast reactor fuel cycle, namely between doubling time and the power output, were not perceived. In 1974 the US AEC published its elaborate cost–benefit analysis, with the following comment from the Environmental Protection Agency: 'trade off between LMFBRs with a low breeding ratio but lower costs/kWh versus LMFBRs with higher ratios and higher costs is in our opinion very important for cost–benefit considerations and may also have significant environmental considerations'[11]. The comment was not taken seriously.
(3) The complexity of the reprocessing and refabrication of fast reactor fuel was not grasped until late in the programme. The application of the PUREX process to fast reactor fuel is only in the development stage. The high costs are now becoming apparent.
(4) Increasing concern over proliferation of nuclear weapons has led to novel

scenarios which minimise the plutonium output of breeders, and even argue that they can be used as incinerators overall and not as producers of plutonium.
(5) The recent perception that nuclear power is only a marginal producer of electricity with no proven benefits may have led to a reassessment in the direction of attempting to maximise benefits in the short run. In this case, the trade-off between breeding and power output will be decided in favour of the latter.

It is perhaps noteworthy that although consideration (4) starts from a very different premise than (5), both lead to the same conclusion. Whatever the weight of the above arguments and others may be, one of the results of the present conception of the breeder is that economic benefits are no longer identified with breeding. The implications are of major importance. For 25 years, official UK policy has been predicated on the emergence of the fast reactor, for three reasons. First, it would overcome the severe constraint of uranium supply. Given that thermal reactors are uneconomical users of uranium (light-water reactors in particular) and the poorest producers of the plutonium, this constraint would operate within a time range that itself is a further constraint. Secondly, the fast reactor would reduce the real costs of electricity and make nuclear power the cheapest option. (Today, when research is beginning to question the economics of nuclear power, to sustain this claim is obviously of great importance.) Thirdly, the fast reactor programme has meant (on paper at least) that uranium resources constituted the largest single source of power in the world among the non-renewable fuels.

The presently favoured scenario of the Atomic Energy Authority, as explained in John Surrey's paper, is to build one demonstration commercial reactor, before any programme is decided upon. For a single reactor, the doubling time has no significant effect on the economics. This conforms, therefore, with the current acceptance of long doubling times—but without any perspective for the benefits in the longer term.

The size and rate of growth of a fast reactor programme depend on the size of the plutonium stocks. These are a function of the thermal reactor programme, the size of the plutonium inventories, the out-of-reactor time and the doubling time. As the thermal reactor programme is relatively small and the plutonium balance thereby adversely affected, the other variables become correspondingly more important. In the short run, the out-of-reactor time is undoubtedly the most important, but in the longer run (and especially if the out-of-reactor time is sub-optimal), then the doubling time will be the most important determinant in the success of the fast breeder programme, including the economic benefits that it will realise. Placing the doubling time low in the order of determinants, as present strategies do, may be no more than a recognition of reality. But it is one that raises questions about the perspective for nuclear power, and its implications ought to be more fully elucidated as a part of the task of assessing where nuclear power policy is going.

REFERENCES

1. Gowing, M., *Independence and Deterrence,* Vol. II, chap. 19, The Macmillan Press, 1974
2. Ibid., p. 267

3. Ibid., p. 267
4. *Nucleonics,* p. 59, Sept., 1966
5. Ibid., p. 69
6. Ibid., P. 59
7. Farmer, A. A. and Hunt, H., 'Important Factors in Fast Reactors' Economics', *Nuc. Eng. Int.*, IPC, July, 1978
8. Ibid., p. 43
9. For CEGB projections see Pipe, E. J., IAEA International Conference, Istanbul, 1969; Parliamentary Committee on Science and Technology, Sub-Ctte B, Appendices to Minutes and Proceedings, 1967
10. Hunt, H. and Betteridge, G. E., 'The Economics of Nuclear Power', *Atom,* December, 1978
11. US Atomic Energy Commission Environmental Impact Statement on the LMFBR, 1974
12. Rosenholz, M., *et al.,* 'Fast Breeder in Europe', *European Nuclear Society Conference,* Hamburg, May 1979
13. Aubert, M., Allain, A. and Simon, M., 'Development of Commercial LMFBR Plants after Superphenix', *ANS Trans.*, May 31st, 1979
14. Department of Energy, Energy Commission Paper No. 6
15. Windscale Public Inquiry, BNFL 232
16. Prior, M., this book p. 131–133
17. Häfele, W. and Starr, C., 'A Perspective on Fusion and Fission Breeders', *B.N.E.S.J.,* **13** No. 2, 131–139, April, 1974
18. Marshall, W., 'Nuclear Power and Non-Proliferation', lecture to the Uranium Institute, London, Feb., 1978

Part IV
Political Problems of Fast Reactors

13

Plutonium and Proliferation Problems

Bhupendra Jasani

13.1 INTRODUCTION

Nuclear energy has manifold applications, but the most controversial of all the peaceful applications are the use of nuclear power reactors to generate electricity and the peaceful use of nuclear explosives (PNEs). The controversy arises from the fact that power reactors used for generating electricity produce, as a by-product, plutonium-239, which can be used as an explosive in nuclear weapons. (The other two nuclear explosive materials are uranium-233 and uranium-235.) So far as PNEs are concerned, the problem arises because there is no essential distinction between nuclear explosive devices used for military and those for civilian purposes, and therefore the development of PNEs by any state means that that country acquires the ability to produce nuclear weapons. In the following sections, it is the first problem, that of nuclear power reactors and plutonium, which is discussed.

Concern about the spread of peaceful nuclear technology and plutonium leading to the proliferation of nuclear weapons resulted in the Treaty on the Non-Proliferation of Nuclear Weapons (NPT) which entered into force on 5 March 1970. The essential relevant provisions of the NPT can be summarised as follows:

Under Articles I and II, the nuclear weapon states are committed not to transfer, while the non-nuclear weapon states are under an obligation not to receive, manufacture or otherwise acquire nuclear weapons or other nuclear explosive devices or control them. Article III obligates the non-nuclear weapon states to conclude safeguards agreements with the International Atomic Energy Agency (IAEA) covering all their peaceful nuclear activities to ensure that there is no diversion of nuclear material to the manufacture of nuclear explosives. Article IV states that all parties to the treaty have the right of full exploitation of the use of nuclear energy for peaceful purposes and obligates those parties in a position to cooperate with other countries in developing peaceful nuclear tech-. nology to do so. Article VI commits all parties to pursue negotiations in good faith on effective measures contributing to the cessation of the nuclear arms race at an early date and to nuclear disarmament, including a treaty on general and complete disarmament. And finally, Article VIII.3 of the NPT states that: 'Five

177

years after the entry into force of this Treaty, a conference of Parties to the Treaty shall be held in Geneva, Switzerland, in order to review the operation of this Treaty with a view to assuring that the purposes of the Preamble and the provisions of the Treaty are being realized. At intervals of five years thereafter, a majority of the Parties to the Treaty may obtain, by submitting a proposal to this effect to the Depositary Governments, the convening of further conferences with the same objective of reviewing the operation of the Treaty.'

The first Review Conference was held in Geneva in May 1975. By this time the ideals expressed in Article IV could hardly be said to have materialised. Nuclear economic stakes are so huge that international nuclear dealings are carried out between industrial firms on the basis of ordinary commercial rules, competition and narrow national interest. At present, the trend is for greater use of reactors using enriched uranium fuel (*Figure 13.1*) so that the supply of enriched uranium fuel will become of particular concern for countries which are less developed in nuclear technology. Despite Article IV, the enrichment monopoly of the USA and the USSR, the principal suppliers of enriched uranium, remains unbroken, although this technology could be made available under international control. The obligations of the nuclear weapon parties, as defined in Article VI, remained essentially unfulfilled in 1975 and the situation has not changed much since then. Nor have the anticipated modest steps, such as a comprehensive nuclear test ban treaty and slowing down the arms race at the US-Soviet Strategic Arms Limitation Talks, materialised.

The 1975 Review Conference did not succeed in solving the problems essential to the survival of the NPT. For example, the conference contained no firm under-

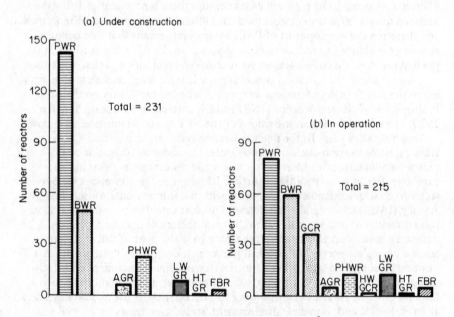

Figure 13.1 Various types of reactors[6].

standing to end discriminatory supplier policies. It emphasised the responsibilities and obligations of all parties, but did not commit the nuclear weapon states to carry out their obligations to reverse the nuclear arms race. The only new outcomes of the Conference were the promotion of international arrangements to ensure the physical protection of nuclear materials and some input into the idea of setting up multinational nuclear fuel cycle centres.

The next Review Conference is scheduled to take place in 1980. What can be done at this conference to strengthen the NPT? Are there any technical solutions which can be introduced to make the nuclear fuel cycle proliferation-resistant or will it be, in the end, a political question?

13.2 WHY SO MUCH CONCERN ABOUT PLUTONIUM?

Of the three fissile materials, only two, uranium-235 and plutonium-239, have been used so far as nuclear explosives. China is the only country which carried out its first nuclear explosion using uranium-235, the other four nuclear weapon powers used plutonium-239 (*Table 13.1*). India also used plutonium-239 in its 1974 underground nuclear explosion. The first nuclear weapon used at Hiroshima in 1945, however, employed highly enriched uranium-235 as the fissile material. Most of the fission weapons in possession of the nuclear weapon powers use ^{239}Pu. Uranium-235 appears to have been used as a fission trigger in some thermonuclear explosions. However, when a B-52 aircraft and a KC-135 refuelling tanker aircraft collided in mid-air in January 1966 near Palomares, Spain, four hydrogen bombs separated from the aircraft and two of these released radioactive material[1]. It has been reported that the radioactive material contained plutonium-239, suggesting that this fissile material can also be used as a trigger in a thermonuclear weapon. At present, therefore, it is clear that plutonium-239 is an important ingredient in nuclear weapons of today.

Until about 1954, most of the information required for the design and construction of both reactors and fission explosives was classified. However, considerable information on how to design a fission bomb is now available in the open literature. To make a number of reliable, efficient and light-weight fission weapons for a national military programme is a complicated business, but a crude

TABLE 13.1 FIRST NUCLEAR EXPLOSIONS (FISSION DEVICES)

Country	Year of first explosion	Fissile material	Source of fissile material
USA	1945	Pu-239	Reactor
USSR	1949	Pu-239	Reactors
UK	1952	Pu-239	Reactors
France	1960	Pu-239	Reactors
China	1964	U-235	Gaseous diffusion
India	1974	Pu-239	Reactor

inefficient nuclear explosive device with unpredictable yield is not beyond the capacity of a small group of people. It may, therefore, be useful to examine the extent to which the other vital part of a fission bomb, the fissile material, particularly plutonium-239, is available, and to look at the measures available to control the fissile material from the point of view of proliferation of nuclear weapons.

At this stage, it may be worth pointing out that, when considering plutonium and proliferation, horizontal proliferation is usually regarded as the main problem. However, plutonium-238 has also a role to play in proliferation issues—in this case, in relation to vertical proliferation. (By vertical proliferation is meant both the increase in number and quality of weapons.) In the following sections, both these aspects are considered briefly.

13.3 PRODUCTION OF PLUTONIUM

13.3.1 Plutonium-239

Any reactor fuelled with uranium produces plutonium as a by-product. The quantity and quality of this plutonium may, however, vary. A power reactor using uranium to generate electricity by the fission process unavoidably produces plutonium which can be used either as a fuel for other reactors or as nuclear weapon material. It has been reported that, in the United States, a fission device was successfully exploded which used reactor-grade plutonium[2]. It follows, therefore, that in order to make an assessment of the proliferation risks, all types of reactors, military as well as those used for power production, must be considered.

A thermal reactor generating 10 MWe produces about 3 kg of high-grade

TABLE 13.2 PLUTONIUM PRODUCTION IN VARIOUS TYPES OF REACTORS[3,4]

Type of reactor	Irradiation level of heavy metal (MWd/kg)	Average enrichment (% U-235)	Initial fuel inventory (kg/MWe)		Pu-239 production (g/MWe annum)
			Core (natural U)	Blanket (depleted U)	
BWR	17	~ 2	434	–	250
PWR	22.6	~ 2.3	365	–	255
AGR	–	~ 1.6	620	–	100
HWR (Candu)	6	0.711	143	–	490
HTGR (USA)	54.5	~ 93	326	–	–
SGHWR	15.5	1.8	520	–	150
FBR (UK)	~ 70	–	9.5(U)*	16.9(U)*	2580 (core Pu)‡
			2.8(Pu)†	–	409 (blanket PU)‡

*Depleted uranium.
†Plutonium composition is 57% Pu-239, 24% Pu-240, 14% Pu-241 and 5% Pu-242.
‡Total of 2980 kg/annum containing 58% Pu-239, 28% Pu-240, 9% Pu-241 and 5% Pu-242.

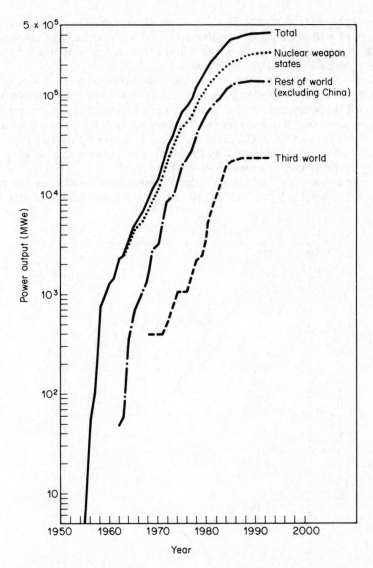

Figure 13.2 Nuclear electrical power generation in various regions[6].

plutonium per annum. From *Figure 13.2* it can be seen that the growth of electricity generated from nuclear energy is expected to level off at about 4×10^5 MWe by about 1990. This means that, by this time, civilian power reactors will be producing some 120 000 kg of plutonium per annum, that is, enough for some 15 000 atomic bombs, each equivalent to 20 kilotonnes of TNT. It is interesting to note (*Figure 13.3*) that some 5% of the total electrical power produced by nuclear reactors will be generated in the Third World countries, excluding China (that is, 6000 kg of plutonium), some 30% in the rest of the world, excluding China (that is, 36 000 kg of plutonium) and some 60% in the four nuclear weapon states, excluding China (that is, 72 000 kg of plutonium).

In *Table 13.2*, plutonium production from various types of reactors is summarised. It can be seen that nearly five times the amount of plutonium produced by the HWR (Candu) is generated in the new type of reactor, the fast breeder. This amount of plutonium, if it is not burned, becomes available from the core of the reactor alone. The significant characteristics of such reactors is that with their conversion ratios of 1.2 or higher, they can produce more fissionable material than they consume while generating electrical power.

The main arguments given to support the development of breeder reactors are the need for national independence and the assertion that any kind of reactor

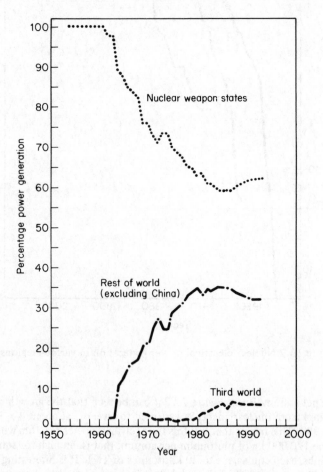

Figure 13.3 Percentage of the total nuclear power generation in various regions.

TABLE 13.3 BREEDER REACTOR PROGRAMMES OF VARIOUS COUNTRIES.

Country	Name and purpose of reactor	Power production		Date of criticality	Status
		Thermal (10⁶ GW)	Electrical (10⁶ GW)		
Federal Republic of Germany (FRG)	KNK II Demonstration	58	17.8	1971	Operational
	KKW Kalkar Demonstration	762	292	–	Construction started in 1973; operation in 1983–8
	SNR-2 Prototype commercial	3418	1300	–	Conceptual design
France	Phenix Demonstration	591	250	1973	Operating
	Superphenix Prototype commercial	3000	1200	–	Construction started in 1977; operation in 1983
Italy	PEC Irradiation testing	140	–	–	Under construction
Japan	Toyo Irradiation testing	100	–	1976	Operating
	Monju Demonstration	714	250	1983	Construction to start in 1979
UK	PFR Demonstration	600	230	1974	Operational
	CDFR Prototype commercial	3230	1250	–	Preliminary design
USA	FFTF Irradiation testing	400	–	1979	Under construction
	Clinch River Demonstration	975	350	1982	Detailed design
USSR	BOR-60	60	11	–	Operational
	BN-350 Demonstration electrical-desalination	1000	135	1972	Operational
	BN-600 Early evolution electrical	1500	600	–	Under construction
	BN-1600	–	1600	–	Planned

presents a proliferation problem, and that breeder reactors are no more proliferation-prone than other types. However, concern about breeder reactors is directed at the fact that the 'plutonium economy', of which breeder reactors form a part, assumes possession both of reprocessing plants, which can separate the plutonium from the spent fuel, as well as plutonium fuel fabrication plants. Moreover, large quantities of plutonium in relatively accessible form are also available in the breeder reactor fuel cycle, which makes diversion of plutonium more feasible.

The fast breeder reactor programmes of various countries are shown in *Table 13.3.*

13.4 CAN PROLIFERATION BE PREVENTED?

The issue of proliferation is a complex one, since a number of questions—technological, economical, environmental and political—have to be considered. Several technical solutions have been proposed to make the present nuclear fuel cycle proliferation-resistant. One such solution is based on the use of power reactors currently operating on a once-through fuel cycle. The idea here is to leave the spent fuel unreprocessed and to store it while it is still in the fuel elements. However, plutonium remains relatively inaccessible only so long as the fuel elements contain very high levels of radioactivity, at which stage, plutonium can only be removed in a sophisticated, expertly operated, well designed reprocessing plant. As the level of radioactivity decreases, however, the degree of sophistication needed in handling the fuel elements also declines. If the plutonium is in the form of plutonium metal or oxide, after reprocessing it becomes available for use in a weapon. If it is mixed with uranium, a further step is needed to separate it from the uranium. Thus, the idea of the once-through cycle is not enough to prevent proliferation.

The second concept is to use denatured fissile materials. In the case of uranium, it has been suggested that the fuel can be a mixture of uranium and thorium. Most of the new fissile material in such a mixture is uranium-233 with some plutonium produced from uranium-238. Similarly, a number of concepts for denaturing plutonium consist of using plutonium isotopes. However, reactor-produced plutonium-238 and plutonium-240 are not effective. Their fission properties are such as to increase the critical mass of plutonium-239 only moderately. Another idea which has been discussed is to mix the element beryllium with the plutonium so as to increase the neutron flux, thus making the safe handling of plutonium difficult. Such a mixture may deter a non-governmental group from acquiring plutonium, but an organisation with more sophisticated facilities can separate out chemically the plutonium from beryllium.

The third concept is to develop fast breeder reactors and to use plutonium generated in the thermal reactors to fuel fast breeders. The idea of recycling plutonium in thermal reactors is less attractive, since such reactors will produce more plutonium, whereas fast breeders can be operated in such a way as to breed just enough plutonium to fuel fast reactors.

Recently it has been suggested that the breeder reactor fuel cycle using plutonium could be made proliferation-resistant by reprocessing the spent fuel in such a way that the plutonium is mixed with a considerable amount of short-lived

184

radioactive fission products present in the spent fuel. In this method, the Civex process, the reprocessed fuel would be a mixture of uranium oxide and plutonium, the concentration of the latter being less than 25%. It must be emphasised, however, that the Civex process is designed basically for reprocessing fuels from fast breeder reactors so that the technique could not be used for some decades.

Moreover, although schemes such as Civex may make it difficult for a small group of people, or even a country with a less developed technological base, to divert plutonium, it is always possible for a sophisticated organisation to extract weapons-grade plutonium if it is really determined to do so. No technical measures can solve such proliferation dangers.

13.5 THE PROBLEM OF PLUTONIUM-238

Generation of heat on a smaller scale can be realised using radioactive sources. A suitable radionuclide is one which has large specific power*, nuclear radiation in the form of alpha emission, a long half-life and relatively inexpensive. Most of these criteria are fulfilled by plutonium-238 (*Table 13.4*).

TABLE 13.4 PHYSICAL PROPERTIES OF SOME RADIONUCLIDE FUELS[5]

Radionuclide	Half-life (years)	Power density (W/g)	Form of fuel	Melting point of fuel compound (°C)	Main radiation
Cobalt-60	5.24	15.8	Metal	1480	β, strong γ
Strontium-90	28	1.0	$SrTiO_3$	1910	β, bremsstrahlung
Caesium-137	30	0.22	CsCl	646	β, a few γ
Promethium-147	2.6	1.8	Pm_2O_3	2130	β, a few γ
Thulium-170	0.35	9.6	Tm_2O_3	2375	α
Polonium-210	0.38	45	PePo	1400–2200	α
Plutonium-238	87.6	2.6–4.0	PuO_2 or PuPo	2250	α
Curium-244	18.1	13	Cm_2O_3	2000–2200	α

Plutonium-238 in a relatively pure form, that is, free from other plutonium radionuclides, is produced either by neutron irradiation of neptunium-237 in a nuclear reactor or by the production of curium-242 in a nuclear reactor.

13.5.1 Proliferation Implication of Plutonium-238

The use of plutonium-238 has implications for the vertical proliferation of nuclear weapons through its use as a power source in artificial Earth satellites. We now know that artificial Earth satellites circling round the Earth in near and geosynchronous orbits are being increasingly used to enhance the performance of Earth-

*Specific power = (liberated energy) ÷ (mass of radioactive material).

TABLE 13.5 PLUTONIUM-238 POWER GENERATORS ON SATELLITES

Satellite	Date of launch	SNAP No.	Power (We)	Total weight (kg)	Comments
US Navy Transit-4A (1961-01)	29 June 1961	SNAP-3	2.7	3.1	Test for developing an integrated navigation system; first nuclear power supply
US Navy Transit-4B (1961-AH1)	15 Nov. 1961	SNAP-3	2.7	3.1	Similar to Transit-4A, SNAP-3, lifetime 8 months
US Air Force/US Navy (1963-38B)	28 Sept. 1963	SNAP-9	–	–	Navigation satellite
US Air Force/US Navy (1963-49B)	5 Dec. 1963	SNAP-9	–	–	Navigation satellite
US Navy Navigation Satellite	21 Apr. 1964	SNAP-9	25	(1 kg of Pu-238)	Satellite failed to orbit; about 17 kCi of Pu-238 were distributed at about 50 km altitude; by 1970 about 95% of this was deposited on Earth's surface
NASA Nimbus 2 Weather Satellite	19 May 1968	SNAP-19	25	14	Two power units were carried by the satellite but guidance malfunctioned and the satellite was exploded; power units recovered
NASA Nimbus 3 Weather Satellite	14 Apr. 1969	SNAP-19	30	13.6	Two power units were carried by the satellite
NASA Apollo 11 Lunar Module (1969-59C)	16 July 1969	SNAP	15 (Th)	–	Early Apollo Scientific Experiment Package was kept warm during lunar night by two Pu-238 power sources
NASA Apollo 12 Lunar Module (1969-99C)	14 Nov. 1969	SNAP-27	63.5	30.9	Apollo Lunar Surface Experiment Package

Spacecraft	Date	Power source			Remarks
NASA Apollo 13 Lunar Module (1970-29C)	11 Apr. 1970	SNAP-27	63.5	30.9 (3.8 kg of Pu-238; 44.5 kCi)	The power source from the Lunar module was jettisoned in the South Pacific Ocean; no contamination was found
NASA Apollo 15 Lunar Module (1971-63C)	26 July 1971	SNAP-27	–	–	Lunar module landed on the Moon on 30 July 1971
NASA Pioneer-10 (1972-12A)	3 Mar. 1972	–	30	13.2	RTG, unmanned spacecraft flew by Jupiter in December 1973
NASA Apollo 16 Lunar Module (1972-31C)	16 Apr. 1972	SNAP-27	–	–	–
US Air Force Triad-01-1X Transit Navigation (1972-69A)	2 Sept. 1972	–	30	13.2	RTG power generator
NASA Apollo 17 Lunar Module (1972-96C)	7 Dec. 1972	SNAP-27	–	–	–
NASA Pioneer-11 (1973-19A)	6 Apr. 1973	–	30	13.2	Spacecraft flew by Jupiter in December 1974 and encountered Saturn in September 1979
NASA Viking-1 Lander (1975-75G)	20 Aug. 1975	–	35	15.9	RTG; Lander landed on Mars on 20 July 1976
NASA Viking-2 Lander (1975-83C)	9 Sept. 1975	–	35	19.9	Lander landed on Mars on 3 September 1976
US Air Force Les-8 (1976-23A)	15 Mar. 1976	–	145	34.1	RTG power generator
US Air Force (1976-23B)	15 Mar. 1976	–	145	34.1	RTG power generator

SNAP: Space Nuclear Auxiliary Power
RTG: Radioisotope Thermionic Generator

bound weapons carrying their lethal nuclear warheads. The increased accuracy of delivery of these weapons to their targets with the aid of satellites may not only provide incentives to development of new types of nuclear weapons but is also contributing to a change in war fighting doctrines, such as the current counter-force and flexible response doctrines, which would emphasise limited nuclear war-fighting capabilities at various levels.

It is indicative that satellites orbited at the beginning of the space age in 1957 needed no more than a watt or two of power to transmit their sensor readings back to Earth by radio, while today's satellites need more power because they perform many complex tasks. Moreover, present development in the anti-satellite systems make the possibility of space warfare more likely. This, in turn, may give impetus to the development of nuclear power generators, for example, to replace solar panels, thus increasing the survivability of satellites by reducing their radar cross sections.

At present, only plutonium-238 nuclear power sources are being used in the United States on military satellites such as navigation, meteorological and communications satellites (*Table 13.5*). However, account should also be taken of the nuclear reactors used to power ships above and below the ocean surface. In future, no doubt, aircraft, missiles and satellites launchers may also use nuclear reactors as propellants. At the moment, these activities only contribute to vertical proliferation, but eventually they may also affect horizontal proliferation.

13.6 CONCLUSIONS

Plutonium is the first man-made element in this nuclear age of ours which has attracted the most discussion and concern. This is because of its dual nature. Plutonium-239 is one of the three fissile materials which can be used both for producing electrical energy from civilian nuclear reactors and for creating nuclear weapons. A wide use of plutonium in the first application has not yet taken place but the increasing use of nuclear reactors will make plutonium available to many national and sub-national organisations. On the other hand, the use of plutonium for weapon production is confined to the five nuclear weapon powers only.

Considerable effort has been devoted to preventing the use of plutonium for weapon production by more nations than the five nuclear weapon powers. As seen above, several technical suggestions have been put forward. However, it seems that the technical constraints on the nuclear fuel cycle cannot prevent a nation determined to acquire nuclear weapons from doing so. At best, it can only make it difficult to divert plutonium from the civil nuclear fuel cycle.

In 1977, the International Nuclear Fuel Cycle Evaluation (INFCE) was set up to make a global evaluation of the comparative merits or drawbacks of the different nuclear fuel cycles and processes from the point of view of their risks of proliferation. The findings of the INFCE are not yet complete.

In the absence of a completely technical solution, other solutions such as limitations or denials of transfers of nuclear facilities, materials and technology are being devised. Examples of these are the recent US legislation for non-proliferation, and the formation of the nuclear suppliers club, the so-called London Club. In the long run, however, such solutions may not work. To avert the possibility of diversion of fissile materials suitable for weapon production from civilian power reactors, the non-nuclear weapon state parties to the NPT are subject to international verification and on-site inspection by the International Atomic Energy Agency (IAEA). These states are obliged to recognise full-scope safeguards. However, with a future increase in the production of plutonium, safeguards technology in its present state may be inadequate to detect any diversion of plutonium. In any case, the function of the safeguards is to detect such diversion in time for an effective response but not to prevent any diversion. It may be in place here to suggest that safeguards should be strengthened.

Another suggestion is to make international the sensitive parts of the nuclear fuel cycle, such as the enrichment and reprocessing facilities. Such a solution is not without its problems either, however. The solution to the problems of proliferation must be both technical and political and has to be connected to progress in nuclear disarmament.

REFERENCES

1. Stockholm International Peace Research Institute, 'Accidents of nuclear weapon systems', *SIPRI Yearbook of World Armaments and Disarmament, 1977*, Stockholm, Almqvist & Wiksell, p. 66 (1977)
2. 'US exploded bomb made from power reactor plutonium', *Nuclear Engineering International*, Vol. 22, No. 263, October 1977, p. 4
3. Marshall, W., 'Nuclear power and the proliferation issue', *Atom*, No. 258, April 1978, p. 101
4. *Nuclear proliferation factbook*, US Library of Congress, 23 September 1977, p. 382
5. Cortiss, W. R. and Mead, R. L., *Power from radioisotopes,* An Understanding the Atom Series Booklet, US Atomic Energy Commission (1971)
6. International Atomic Energy Agency, *Power Reactors in Member States*, IAEA (1978)

14

Nuclear Power and Civil Liberties

David Widdicombe, QC

14.1 INTRODUCTION

Of the various objections to the use, or extended use, of nuclear power as a source of energy supply (safety, disposal of wastes, radiological hazards, etc.), the one I am concerned with in this chapter, namely the impact of the nuclear programme on civil liberties and the democratic way of life, may well be the most important. The Royal Commission described it as 'a central issue' in the debate over the future of nuclear power (para. 186).

The subject is now well documented. The Royal Commission on Environmental Pollution Sixth Report deals exhaustively and extremely clearly with the general issues (para. 182-186, 305-336) and more detailed studies exist in the JUSTICE Report, *Plutonium and Liberty*, and the Friends of the Earth publication, *Nuclear Prospects*. There is a wealth of American material, both official and unofficial.

It is not possible in this short paper to set out all the facts and arguments covered by these documents. Nor is it necessary to do so, because to a large extent the basic facts and the main conclusions to be drawn from them are not in dispute. I think it is now common ground that dependence on nuclear power, and particularly plutonium, has inevitable consequences for civil liberties and for democracy, most of which do not appear to be involved in other forms of power production, whether existing or contemplated. The official responses (e.g. the White Paper, cmnd 6820, May 1977, and the Windscale Inquiry Report) do not seek to deny the problem, but concentrate on the argument that dependence on nuclear power has not yet gone so far as to produce these effects, and on suggestions for ameliorating the situation if it arises.

I shall therefore first briefly summarise the dangers of nuclear power, secondly describe the consequences which flow from those dangers and then turn to what I think are the main questions in the debate, namely how far along the nuclear road have we gone, and how far, from a civil liberties point of view, along that road can we afford to travel.

Although this conference is limited to the fast breeder programme, I have quite deliberately not confined my paper to that. It is true that most of the dangers to civil liberties arise, or will arise, from the widespread use of plutonium as a nuclear fuel, and it is the fast breeder which uses plutonium. And it is true that the threat to civil liberties, as the Royal Commission points out (para. 161, 320) arises when plutonium is in widespread use as a staple commodity of energy supply, when it is regarded as 'a routine item of commerce'. But it is my contention that you cannot separate into watertight compartments the various stages of development of nuclear power. Spokesmen for the nuclear industry have consistently pointed out that the reprocessing of spent nuclear fuel to obtain plutonium and the use of that plutonium in fast breeders have been implicit in the use of nuclear power from the start. This is indeed true and it means that if we do not want a plutonium economy, we may need to question the use of nuclear fission power at all. I will develop that argument later. The question which then must be considered is how do you stop a juggernaut which has started to roll? Can you stop it?

My concern is with civil liberties in the UK. But terrorism is international and likely to become more so. The use of plutonium throughout the world and the dangers of proliferation are very relevant to the situation in the UK, but that aspect of the subject is beyond the scope of this paper.

14.2 THE DANGERS OF NUCLEAR POWER

It is a characteristic of nuclear power, distinguishing it from other forms of power production, that it generates power from material which is extremely hazardous to health, and in the case of plutonium, highly explosive also.

Of the health hazard of plutonium, the Royal Commission says, 'The dispersion of a small amount of plutonium into the atmosphere with conventional explosives would pose a very serious radiological hazard since an individual dose of only a few milligrams is sufficient, if inhaled, to cause massive fibrosis of the lungs and death within a few years. Much smaller quantities can cause lung cancer after a latent period of perhaps twenty years' (para. 322).

The explosive properties of plutonium were sufficiently demonstrated to the world at Nagasaki in 1945 to need no comment. That was a government bomb. Since then, the necessary information to make an atomic bomb has been freely published, and their manufacture by persons or groups outside government has become possible. The Royal Commission found that, 'The equipment required would not be significantly more elaborate than that already used by criminal groups engaged in the illicit manufacture of heroin' (para. 323). A large number of people in the world now possess the necessary scientific knowledge and skill to design and make a crude atom bomb—one estimate, I believe, puts it at a million. The Royal Commission concluded that 'it is entirely credible that plutonium in the requisite amount could be made into a crude but very effective weapon that would be transportable in a small vehicle' (para. 325). Since that statement, a report has been published about the use of atomic demolition munitions (ADMs) or 'suitcase atom bombs', so small and portable that they can be carried in knapsacks and fired by only two people, yet are powerful enough to destroy a city block[1]. Several crude atom bombs have, in fact, been made or designed by amateurs, though without fissile material, and have been submitted to nuclear weapons experts, who have been unable to fault them[2]. The cost has been put at about £10 000.

The problem for a terrorist group is not therefore the making of the atom bomb but the obtaining of the fissile material, highly enriched uranium, or plutonium. At one time, it was argued that only 'weapons-grade' plutonium is suitable, but it is now accepted that 'reactor-grade' plutonium is suitable. Six to ten kilograms (the size of a grapefruit) is said to be enough, giving a yield of 100 tons of TNT equivalent, enough to demolish the Houses of Parliament, Westminster Abbey and half the Government offices in Whitehall, if detonated in Parliament Square. Once obtained, because of the nature of its radioactivity (alpha), plutonium is easy to handle and very difficult for security forces to detect.

It should be noted, however, that although it is necessary to have the fissile material in order to make a bomb which will explode, it is not necessary to have it in order to make an atom bomb threat. It may be enough for terrorists to make people believe that they have the material. This is why many people are worried about the impossibility of accounting accurately for the quantity of plutonium involved in the process of separation and manufacture of fuel elements; as in all manufacturing processes, there is material unaccounted for ('MUF'). The cumulative quantity of plutonium unaccounted for at Windscale between 1970 and 1977 was about 96 kg. Uncertainty is increased by the claims, which there is no way of verifying, that there is already an international black market in fissile materials, including plutonium[3]. According to papers recently declassified under the Freedom of Information Act in the United States, 44 nuclear threats credible enough to be taken seriously have already been made against US cities or industrial plants since 1970[4]. The UK Government say that there have been none so far here[5].

14.3 CONSEQUENCES FLOWING FROM THE DANGERS

Certain consequences flow from the use by the nuclear industry of materials which are hazardous to health and highly explosive. Firstly, nuclear installations and materials in transit require special guarding against release of radioactivity and theft of fissile materials, and secondly, security vetting of personnel and surveillance of members of the public are an essential precaution.

14.3.1 Nuclear Installations

It is true that coal- or oil-fired power stations, and indeed solar-, wave- or wind-power installations, could all be sabotaged, but the worst that sabotage can do to such installations is to bring them to a halt. In the case of a nuclear power station, enrichment works or reprocessing plant, in addition, there is the danger that the 'inventory' of nuclear materials could in the whole or in part be released into the atmosphere. The danger of this happening appears to be less in the case of our present 'thermal' nuclear power stations, than in the case of reprocessing plants and fast breeders. There is also the problem that, if in the future we become to a material extent dependent on fast breeders for our electricity supply, it will be necessary to guard reprocessing plants like Windscale with especial care, because if they were put out of action for any appreciable time, the fast breeders might have to close too. Installations using or containing supplies of plutonium require guarding not only against sabotage but against theft. The general picture was summed up by Tony Benn when he said, 'You have to protect plutonium with absolute maximum security. . .Dounreay (the breeder reactor site) was a lovely research centre on the north coast of Scotland in 1966, like Princeton, with pipe-smoking professors. You go back now and it is an armed camp with barbed wire, guard dogs, arc lights and an armed constabulary'[6].

14.3.2 Materials in Transit

The nuclear cycle takes place in a number of widely separated plants, and materials have to be moved all round the country, the number of movements increasing as the industry grows. All modes of transport are used, air, sea, road and rail. Spent fuel, which is highly radioactive, and therefore a target for sabotage, is transported in armoured 'flasks'; fissile materials and fuel rods are not so dangerous to health but have to be guarded, especially plutonium, against theft.

The need for security guarding of nuclear installations and materials in transit has led to the setting up under the Atomic Energy Authority of an armed nuclear police force, authorised to carry and use automatic weapons, to engage in hot pursuit of actual or potential thieves, and to arrest on suspicion[7]. At present, it numbers about 400, and it will grow in size as the nuclear industry grows[8]. Because the Atomic Energy Authority is not a Government department, but an independent statutory corporation, there is no Minister responsible for its day-to-day operations in Parliament, and the police force is largely beyond the scope of democratic control. It has been convincingly argued that the absence of demo-cratic control is inevitable and deliberate because of the security implications of the force's operations[9].

14.3.3 Vetting and Surveillance

Personnel employed at nuclear installations and those engaged in transit operations have to be security 'vetted' either by positive or negative procedures. This is done by the Atomic Energy Authority for personnel at its own, or British Nuclear Fuel, establishments. Vetting, of course, did not prevent the infiltration of our nuclear establishments by agents of Russian Communism during and after the war. It should be noted that Alan Nunn May handed over, not just information and documents, but actual samples of nuclear materials produced here. The possibility of infiltration and theft of materials from the inside, by a political or other motivated organisation, is therefore a real one—it has actually happened. In addition to vetting employees, security surveillance of family, friends and associates may be necessary.

One special problem which causes concern is the need for security surveillance of individuals and organisations opposed to nuclear power. The necessity for this may arise not only after a nuclear threat has taken place, but also before it, if one is anticipated. The Government has stated that, 'Bodies and individuals opposed to the development of nuclear power would not be subject to surveillance unless these was reason to believe that their activities were subversive, violent or otherwise unlawful'[10]. 'Subversive' is not a word known to the English law. It was defined in a Parliamentary answer as follows: 'Subversion is defined as activities threatening the safety or well-being of the state and intended to undermine or overthrow Parliamentary democracy by political or violent means'*. This is very wide. A good argument could be made out, for example, that persons who want to abolish the House of Lords are 'threatening the well-being of the state and intending to undermine Parliamentary democracy'. And who decides whether an individual or body is subversive? Presumably the security forces. There is no democratic control over their operations; Ministers refuse to answer questions in Parliament about them†. It has been pointed out that, as nuclear power expands, the need for surveillance is likely to have a restrictive effect on free discussion of nuclear issues; there will be people who will prefer to conceal their views rather than run the risk of saying what they think and having their names added to a list in some secret office.

As I pointed out earlier, the Government does not deny that all these consequences could flow from a widespread use of plutonium as a source of power. It has, however, been suggested that there may be certain measures which could be taken to ameliorate some of the dangers. For example, nuclear installations could be designed with as much in-built security as possible. However, for security reasons it is not possible to disclose what such measures might be or to discuss how effective they may be. Then it is suggested that the danger of theft of plutonium in transit could be met, firstly by ensuring that it is only transported in fuel rods, i.e. combined with uranium compounds and secondly, by 'spiking' it with radioactive materials. As regards fuel rods, the processes necessary to separate the plutonium are not regarded by scientists as particularly complicated (unlike the processing of spent fuel, which is highly complicated.) 'Spiking' clearly is a useful and important safety precaution. It has yet to be proved feasible; it would mean that fuel rods would have to be remotely handled and transported like spent fuel, and it could prove very costly. It would also afford no protection against irrational terrorists who place no value on their own lives or health.

To sum up at this stage, what we face if nuclear power expands is that a large and increasing part of our electricity supply will have to be produced under quasi-military conditions, conducted by independent statutory corporations largely beyond the reach of democratic control. Any failure or slackening of maximum security, either internal in respect of the individuals involved in the operations, or external in respect of sabotage or theft, could have truly disastrous consequences in terms of human lives and the fabric of our society. Unfortunately, history shows only too clearly that security measures, like everything else, are subject to human error and fallibility. It is difficult to avoid the conclusion to which Sir Brian Flowers has been led, namely that it is not a question of *whether* someone will deliberately acquire plutonium for the purposes of terrorism or blackmail, but only of *when* and *how often*.

All sane persons must surely endorse the major conclusion of the Royal Commission that we should not rely for energy supply on a process that produces such

*Compare Mr Merlyn Rees, Home Secretary: 'The Special Branch collects information on those whom I think cause problems for the State' (*Hansard* 2 March 1978).
†Tony Benn has called for an inquiry into the security services (*Guardian* 5 February 1979).

a hazardous substance as plutonium unless there is no reasonable alternative, and that a major commitment to fission power and a plutonium economy should be postponed as long as possible (para. 506, 507, 522). Some might add that even if there is no reasonable alternative they would rather 'read the Bill of Rights by candle light than not have it to read at all'[11].

The next question to consider, therefore, is how far along the nuclear road have we gone, and where should we call a halt.

14.4   HOW FAR ALONG THE NUCLEAR ROAD HAVE WE GONE?

In the 25 years of the civil nuclear programme in the UK, we have built or are building some 14 nuclear power stations (9 Magnox and 5 AGR), which together will account in 1980/81 for some 15% of our generating capacity and some 20% of electricity consumed. All these stations are continuously manufacturing plutonium. The plutonium from the Magnox stations has been separated by reprocessing from the beginning; the plutonium from the AGRs will now be separated as a result of the approval given to the new Windscale plant. All this plutonium is being stockpiled with a view to use in fast breeders. By the end of the century there will be about 75 tonnes of it*.

The programme for the next 20 years until the end of the twentieth century (see Green Paper on Energy Policy, Cmnd. 7101, February 1978) envisages a stepping-up of the pace of nuclear development. This programme is itself much reduced from the ambitious forecast put by the Atomic Energy Authority before the Royal Commission, but it is still substantial. Based on an annual growth rate of 3%, an additional 35 GW of new nuclear power stations is envisaged, which together with 5 GW of AGR plant already built or building, and still operational in the year 2000, will give a nuclear capacity in that year of 40 GW. It appears that this will constitute some 40% of total electrical generating capacity, and a higher proportion (50%?) of electricity consumed. This represents the construction in the next 20 years of some 20 to 30 new nuclear power stations†. All of these will be manufacturing plutonium, which presumably will be separated by reprocessing. How many of these nuclear power stations will be thermal, and how many fast breeders is not stated, though the document (para. 10.10) appears to contemplate that it would be possible to start a major programme of fast breeders from about 1988, assuming approval of CFR-1 in 1979.

It will be seen, therefore, that those who say that the plutonium economy is already with us are fully justified. Certainly by the year 2000, if the present forecast comes about, we shall have a very substantial dependence for our electricity supply on nuclear power, a large and increasing stockpile of plutonium, and quite possibly a sizable and increasing number of fast breeder reactors. On any reckoning, we shall then surely have arrived at that dependence on plutonium which the Royal Commission took as the hallmark of the plutonium economy.

---

*The total plutonium requirement for a fast breeder (1 GW) is about 5 tonnes.
†The number depends on the size of the stations. Finding the sites is not mentioned! The present Government is talking of 15 GW in the decade from 1982.

14.5 WHERE SHOULD WE CALL A HALT?

In considering this question, it is important to appreciate that the growth of the nuclear industry is a continuous process which cannot be broken down into separate and isolated stages. In particular, it is clear that there is no distinct dividing line between the thermal phase of nuclear power and the fast breeder stage. The industry itself has always regarded the fast breeder as the inevitable development; thermal reactors produce plutonium which is then used in fast breeders. Much is made of the difficulty of keeping spent fuel, and how wasteful it would be if the plutonium which has been manufactured is 'thrown away'. As each step is taken, it provides arguments for the next.

Furthermore, as nuclear power grows, the industry itself becomes an ever more powerful vested interest group. The number of employees directly or indirectly dependent for their jobs and future prospects increases. Expansion is self-fuelling. There is a need 'to keep the team in being', 'to provide work for the construction industry', 'to safeguard employment', etc. Nuclear power is not just electricity, it is power in the political sense for the nuclear industry. Already the nuclear industry has all the symptoms of a powerful interest group—a splendid advocate in the role of Chairman (Sir John Hill), extremely ambitious, indeed almost megalomanic, programmes of expansion, and (one suspects) a strong foothold in the Government department responsible for its future.

This is why it was naive for the Windscale Report to find that the THORP reprocessing plant would have no significant effect on civil liberties (para. 7.11). If the THORP plant is taken as an isolated step, the conclusion can perhaps be justified. But if THORP is taken in its context of the development of the nuclear industry over a period, it clearly has significant possible consequences for civil liberties, which should have been recognised. No single step in the movement towards dependence on nuclear power ever will by itself involve a substantial interference with civil liberties. The erosion will be gradual, and it is vital to recognise that fact.

One is inevitably led to the conclusion, with the benefit of hindsight, that we should never have embarked on nuclear power at all. Having done so, we must surely now call a halt as soon as possible. Windscale was a missed opportunity. It will be more difficult to stop at the next step, the CFR-1. But if civil liberties matter to us, that is where we should stop.

REFERENCES

1. *Evening Standard*, 27 September 1978
2. In the UK, see: *Daily Express,* 17 May 1977 (Dr Kit Pedlar and others); Dr Pedlar gave evidence at the Windscale Inquiry about how to make one. Also *Guardian,* 13 June 1977 (group of science undergraduates). Sir Brian Flowers has said that they could make one at Imperial College.
3. For example, *Rolling Stone*, 19 May 1977
4. *Evening Standard*, 17 September 1978
5. Replies by Secretary of State for Energy to Questions by Friends of the Earth, 2 June 1977

6. *Sunday Times*, 20 March 1977
   Vole, March 1977, p. 10
7. Atomic Energy Authority (Special Constables) Act 1976.
   The guarding of materials verges on farce. It is reported that in the USA nuclear waste is transported in vehicles disguised as holiday caravans with guards each of whom carries a .357 magnum pistol, a 12-gauge shotgun, two M16 rifles and a grenade launcher.
   Evening Standard 3 August 1978
8. *Hansard*, 23 February 1976, col. 18
9. *Nuclear Prospects*, pp. 22–3
10. Replies by Secretary of State for Energy to Questions by Friends of the Earth, 2 June 1977
11. Comey, D. D., 'The Perfect Trojan Horse', *Bulletin of the Atomic Scientists*, June 1976

198

Appendices

Appendices

Appendix 1

UK Energy Needs and Energy Supply*

UK PROSPECTS AND STRATEGY

Short and Medium Term

Many of the investment and other decisions which will determine the supply of
energy up to 1985 have already been taken. It will not be easy to vary the volume
and pattern of supplies up to that date, though it will become progressively less
difficult from then on.

For some years ahead we are likely to have, in total, more capacity than we
immediately need. Installed capacity for electricity generation in the UK was in
total some 37% greater than maximum demand in 1976/77, though the margin
varied considerably between the different supply systems. This is significantly in
excess of the normal planning margin provided to ensure security of supply in the
event of severe weather or plant breakdown. Since there is some 15 GW of new
plant already under construction, and due to be commissioned by 1980 or shortly
thereafter, it is expected to be some years before this excess is taken up by rising
demand. Natural gas supplies are forecast to rise by about 40% between 1975 and
1980 as new Northern Basin developments come on-stream, and a further rise is
expected in the mid-1980s.

Several problems will or could arise as a result of this temporary abundance of
supplies. Chapter 9 [of the Energy Policy Consultative Document] has discussed
the conflict between the lack of need, on demand ground, for early orders for new
generating capacity and, on the other hand, the need to maintain a viable plant
industry, and a nuclear design and construction capability, to meet the foreseeable
needs of the later 1980s and 1990s. If nuclear power is required to meet a substantially increased share of our energy needs towards the end of the century, the
market for electricity will have to grow. If demand were held down too long by
competition from cheap and abundant gas, it might be difficult for the market for
electricity to grow sufficiently quickly to absorb whatever increase in supplies was

---

*Section III of the Energy Policy Consultative Document, cmnd 7101, Department of Energy,
pp. 70–80, 111–112.

needed to balance primary fuel supply and demand. The extent and timing of any such switch are, however, at present too uncertain to call for any immediate action to stimulate electricity demand.

The coal industry could also have short-term problems. Production cannot be varied rapidly and the industry must maintain adequate capacity and manpower on the base of which it can expand to cope with increased demand in the long run. Meanwhile, in the short-term, abundant supplies of other fuels combined with low total energy demand may affect the market for coal. If temporary surpluses should emerge, there are a number of measures which can be taken. The Government promised in the 1974 Coal Industry Examination that they would consider these if necessary, and the 1977 Coal Industry Act ensures that the necessary legislative framework is available. It includes powers:

(a) To give stocking aid to help the NCB maintain stocks of coal at the pithead or with certain major customers.
(b) To assist extra coalburn by the electricity boards. The Government has agreed to make available up to £7 million a year for five years to enable the NCB and the South of Scotland Electricity Board to enter into a coal supply agreement. It has also agreed to assist coalburn in Welsh power stations.

Other possible measures to deal with temporary surpluses include:

(a) Additional exports of coal to Europe;
(b) Modification of the opencast programme. This, however, could have a serious impact on NCB finances since opencast coal is the most profitable section of the business.

There is no need for further action at the moment in addition to that already taken. But the situation will be kept under review, in the light of the needs of the industry and of constraints on public expenditure.

It is important that any action taken to deal with immediate short-term problems is consistent with the policies we need to pursue for securing our long-term needs, to which we now turn.

The Prospects for the Longer Term

Any survey of the longer-term future has to take into account assumptions, or a range of assumptions, about the principal factors which will affect the sort of energy policy that is required but which lie outside the range of events which energy policy itself can substantially influence. Such factors include the state of the world energy market, especially the price and availability of oil, and the rate of growth of the UK economy. The world background has been discussed in Chapter 3 [of the Energy Policy Consultative Document].

Economic Growth

The main influence on energy demand is likely to continue to be the rate of economic growth. The current energy forecasts . . . are related to two alternative

202

assumptions about future growth of the UK economy. One represents broadly a continuation of past trends, but takes account also of the hope that North Sea oil will, in the 1980s, reduce somewhat past constraints on growth. This assumption corresponds broadly to a growth rate of 3% per annum, tending to flatten out beyond the end of the century. The forecasts examine also a lower rate of growth, falling to an annual rate of less than 2% by the end of the century. Alternative self-consistent pictures of the economy have been built up over the period, taking assumptions not only about overall growth but also about other key variables, such as the index of industrial production and consumers' expenditure.

There can be no certainty that these two alternative assumptions, the difference between which in 2000 is equal to about half our GDP today, range widely enough to embrace the course that events will actually take. It can be argued that the past two or three decades, which have seen rapid growth in the industrialised world as a whole, though much less so in the UK, were exceptional and that hopes of maintaining or even improving that rate of growth in future are illusory. However, the desire for higher material living standards is very widespread and the policies of almost all Governments are aimed at promoting them. In the UK, as elsewhere, a main aim of energy policy must be to ensure that lack of energy does not frustrate this aim. We need, therefore, to plan so that we are able to meet, if necessary, the energy demand stemming from the higher rate of growth assumed here, that is a rate only a little higher than we have previously achieved.

Energy Demand

The methods currently used by the Department of Energy to forecast energy demand have been described [in the Energy Policy Consultative Document]. Any forecast has to start from an analysis of past trends, but these are not projected automatically into the future. Where there are grounds for believing that past trends are unlikely to continue in the future, an appropriate allowance is made in the forecasts, for example in the assumption that the past rapid increase in domestic central heating will approach saturation within the forecasting period.

The demand forecasts thus start from trend projections derived from past relationships between energy demand and factors such as consumers' expenditure and industrial production; such projections assume continuing technological change of the kind that has in the past affected the ratios between energy demand and economic activity. In addition the forecasts make, in most of the sectors, substantial allowances for the effects of past and expected future price increases. These allowances are equivalent to a reduction in primary demand of about 100 mtce in the year 2000.

We cannot rely on these savings accruing automatically, simply as a result of higher prices. To reinforce the price message and make it fully effective, efforts to conserve energy will have to be intensified and sustained at a high level. With a maximum effort it might be possible to achieve even larger savings, though it is not possible to say, at this stage, what further measures would be required. The savings achievable through active Government conservation policies illustrate energy conservation's place as an 'extra energy source' in our energy economy.

But energy conservation, however vigorously pursued, will not prevent a growth in energy needs. An improved ratio of energy consumption to economic activity

203

can be achieved while past inefficiencies are being corrected. But the scope for such improvement is limited. In the longer term, economic growth of the traditional kind will inevitably lead to growth in energy consumption too.

Table A1.1 shows the forecasts of energy demand, up to 2000, on the two alternative assumptions about the future growth of the economy. The range of forecast consumption in 2000 is significantly lower than that shown in the Energy Policy Review. The upper end of the range shown in the Review was based on a scenario which included the assumption that oil prices would not rise in real terms.

TABLE A1.1 UK ENERGY DEMAND 1975–1985–2000
Values are million tons of coal equivalent.

	1975	1985	2000
Higher growth			
Energy	314	375	490
Non-energy and bunkers	27	40	70
Primary fuel demand	341	415	560
Lower growth			
Energy	314	350	390
Non-energy and bunkers	27	40	60
Primary fuel demand	341	390	450

This is no longer seen as probable. The current lower forecasts of primary energy demand also reflect a lower electrical component in final demand (and hence lower conversion losses) and a larger allowance for conservation.

Energy Supply

[The] contributions which the various energy sources could be making towards meeting the demand for primary fuel in 2000 [are] shown in the table. They are [now reviewed]:

Coal. The NCB is aiming to produce 150 million tons of deep-mined coal in 2000, together with 20 million tons of opencast. It is not possible to be sure how speedily new capacity can be introduced, and these objectives may not be fully achieved. If UK supplies fell short of demand, we might have the option to import coal, though the availability and cost of coal in world markets at that time are uncertain.

Nuclear. A prudent view of the maximum nuclear contribution that can be provided in 2000 is about 95 mtce if it is competitive in price. If no nuclear power plant were built after the completion of the reactors currently under construction and those for which orders are being placed, the nuclear contribution would be about 25 mtce at that time.

Natural Gas. If the remaining reserves available to the BGC do not exceed 55 tcf, natural gas would probably not contribute more than 50 mtce in 2000 to our primary energy. If 75 tcf were available, natural gas might contribute as much as 90 mtce, and the decline would be postponed for a few years, until early in the next century. But this higher availability could include substantial imports from the Norwegian Continental Shelf or elsewhere, which would be a burden on the balance of payments.

Oil. On current assumptions about the UK's total recoverable oil reserves and about the production profile, indigenous oil production would fall to about 150 mtce by 2000. Production could be lower if future discoveries failed to bring reserves up to the assumed total of 3½ billion tonnes, or if production rose to a sharper peak in the 1980s. If reserves turned out to be nearer the upper end of the assumed range (4½ billion tonnes), indigenous production in 2000 could be significantly higher.

Renewable Sources. As yet there are no technical or economic data on which to base forecasts of the contribution from renewable resources in 2000. The range of 30-40 mtce . . . is the sum of the total contributions that might be technically feasible under each of the possible sources. As such, it cannot be treated as assured. It would be more realistic to assume total production in 2000 of 10 mtce at most. The contribution could, however, increase steadily from then on.

Our indigenous energy supplies in the year 2000 might then comprise:

Coal	170 mtce
Nuclear/hydro	95 mtce
Natural gas	50-90 mtce
Indigenous oil	150 mtce
Renewable sources	10 mtce
	475-515 mtce

If the higher growth examined [above] materialises, we should not be able to meet demand from indigenous sources and should need to be importing some 45-85 mtce of primary fuel a year, in addition to any gas imports included at the upper end of the range of gas availability. Such an import requirement might appear quite supportable, even allowing for the assumption that the real price of oil in international markets will then be at least double its present level. Account must, however, be taken of the following facts:

(i) The production figures, particularly for coal and nuclear, are upper limits and will not be achieved without very great efforts. The import requirement would be very greatly increased if production in either or both these industries fell substantially short.
(ii) The availability of indigenous natural gas and oil in 2000 depends on future exploration and exploitation; while the Government can influence the effort put into the search by the conditions of licensing and taxation, the basic uncertainty is how much oil and gas are there to be found.
(iii) Even at this level the contribution from renewable resources is uncertain.
(iv) Though we cannot postulate any particular limit to the volume of inter-

nationally traded oil that might be physically available to the UK early next century, we cannot assume that imported oil (or even imported oil and coal taken together) will necessarily be available, as in the past, to fill any deficiency. We must expect that the energy demands of the third world, including those of the OPEC members themselves, will be rising strongly, adding to the pressures on the world's available fossil fuel reserves.

(v) Without substantially enhanced efforts to conserve energy, demand would be likely to be even higher than here assumed.

Determinants of a Long-term Strategy

The key points of which an energy strategy must take account are:

(i) The lead times for making substantial changes in the pattern of energy supply and demand, and particularly for introducing new technologies, are very long.

(ii) Early action will therefore be needed to deal with difficulties thought likely to emerge in 10 or 15 years time or later.

(iii) The uncertainties in planning so far into the future are very great.

(iv) The consequences for the world of a severe physical shortage of energy are unpredictable, but could be very grave.

(v) To guard against such risks and to ensure future supplies, it is worth paying substantial insurance premiums.

(vi) The long-term energy problem is a world problem. Wide-ranging and effective international collaboration is essential.

In the light of these factors, there can be no question of constructing a blueprint which predetermines all the decisions required over the next 10 or 20 years. The aim must be rather to ensure that present decisions are pointing us in the direction in which, having regard to all the uncertainties of the long-term future, it seems desirable that we should for the present be moving. Future decisions, taken in the light of later and presumably better information, can if necessary bring about a change of course, though not necessarily quickly. But in order to preserve and develop the options we may wish to exercise five, ten or more years hence, we need to take some decisions now so that the necessary technologies and manufacturing capacity are available. But because there is a limit to the insurance premiums we can afford to pay, we cannot pursue a policy of keeping open all options all the time. Flexibility cannot be obtained by postponing decisions, since a decision postponed can mean an option closed, or at best an option that may not be available when required.

No policy can be guaranteed to be equally appropriate for all the possible future circumstances which could arise. Therefore, rather than selecting a policy which would be best in one possible future situation, but which might turn out to be expensive or even disastrous in others, it is preferable to act in ways which would be most beneficial or least costly over a wide range of possible outcomes. Such courses of action may be described as 'robust' or as producing 'minimum regret'. The selected policies need to be continually reviewed and updated, to take account of changing circumstances.

We need to ensure that our policies not only enable us to meet foreseeable demands in the period up to 2000 but also put us in a position to meet energy needs in the twenty-first century. Uncertainties inevitably multiply as one looks further ahead. We do not know what technologies will be available, what life styles will be or what will be society's expectations in the twenty-first century. Most societies that have enjoyed economic growth have wanted steadily increasing material living standards and this has meant increases in energy demand. With fuels scarce and increasingly expensive, the previous relationship between economic activity and energy demand is likely to change, perhaps substantially, but it seems reasonable to assume that demand will continue to grow. We must have the potential to meet that demand, in a period when indigenous production of oil is declining, perhaps rapidly, and when indigenous production of gas is declining or is about to decline.

As we move into the next century, the world's available oil will need to be increasingly reserved for uses for which other fuels cannot readily be substituted, particularly in transport and for petrochemical uses. To some extent coal may be able to fill the gap caused by the withdrawal of oil from crude heat production, but coal itself may be increasingly needed as a raw material, for the production of substitute natural gas, of transport fuels and perhaps of petrochemicals, and the amount available for electricity generation may decline. One possible path for the world's, and the UK's, energy economy in the next century is that it will become increasingly reliant on nuclear power. Fusion power and renewable resources such as solar and wind power are sometimes suggested as an alternative, and environmentally more acceptable, ways of meeting our energy needs. But we cannot be sure at this stage when fusion power will prove practicable or even that it will prove practicable at all. The contribution from renewables is also uncertain. At present the only assured source in the long term, apart from coal, is nuclear power. Dependence on nuclear power, because of the limited availability of uranium, almost certainly involves, before the end of the first quarter of the next century, major reliance on fast reactors. We cannot be certain that, in the absence of fast reactors, alternative ways can be found to meet the world's energy needs in the first quarter of the next century, in sufficient quantity and at an acceptable cost. It will be for society to take the eventual decision whether it is prepared to accept fast reactors or whether, if there is no suitable alternative, it is prepared to accept lower national living standards as a result either of diversion of resources into very high-cost forms of energy or of energy shortages. While there can be no certainty at this stage that fast reactors will be needed, it is probable that they will be. We need, therefore, to keep open the option of fast reactors while at the same time continuing with the development of fusion power and putting increased resources into the development of renewable forms of energy.

An Alternative Strategy

The Royal Commission on Environmental Pollution considered that, because of the possible risks involved in a large-scale reliance on plutonium, a large programme of fast reactors should not be adopted unless there is no reasonable alternative. They suggested an alternative energy strategy for the UK, by which they sought to show that the UK's forecast demand for energy could be satisfied without reliance either on the fast reactor or on imported oil. They described it as an alternative strategy, because they contrasted it with an 'official strategy' deduced from supply and demand forecasts included in the evidence submitted to them. As has been explained in the White Paper replying to the Royal Commission's report, the Government believe that this contrast may have been based on a misinterpretation of the Government's energy policy, which does not set out to provide a blue-print for the development of each form of energy over long periods.

The Commission's alternative strategy, which is examined in more detail [below], includes a substantial thermal nuclear element (50 GW in 2010, 80 GW in 2025) but does not provide for fast reactors. The Commission did not put forward this strategy as a seriously studied option. The analysis . . . suggests that it may well be overoptimistic in a number of respects in its estimates of contributions from non-nuclear sources. It would not therefore be safe to rely on this strategy to meet the UK's energy needs in the period up to 2025. Nor, for the reasons explained. . . would it be prudent to close the option of fast reactors.

THE ROYAL COMMISSION ON ENVIRONMENTAL POLLUTION: AN ALTERNATIVE STRATEGY

The Royal Commission on Environmental Pollution, in their Sixth Report, put forward an 'alternative' long-term strategy for the UK in which they sought to show that reliance on the fast reactor or imported oil in the twenty-first century could be avoided. The Commission's alternative strategy included four main elements:

(a) a substantial thermal nuclear element (50 GW in 2010, rising to 80 GW in 2025);
(b) a heavy reliance on coal, waste materials, winds, waves, tides and sun as primary energy sources;
(c) minimising energy loss in electricity generation by installing combined heat and power systems on a major scale;
(d) matching types of heat required to demand, for example, utilising low-grade heat for space heating and reserving high-grade electricity for premium uses.

The Commission's 'alternative strategy' was put forward not as a seriously studied option but as a contribution to the public debate on energy strategy for the future, focusing attention on the contribution the alternative sources could make and on the more efficient use of energy through reducing losses in production and distribution and matching types of heat to market requirements more closely.

The Royal Commission have proceeded by considering the level of demand for energy to 2025 and considering how it might be met.

Up to 2000

Up to the year 2000—the time horizon considered in detail [in the Energy Policy Consultative Document] —the Royal Commission forecasts of final demand differ only marginally from the Department's forecasts [see *Table A1.2*].

TABLE A1.2 COMPARISON OF FORECASTS OF FINAL DEMAND (EPCD—figures
given by the Department of Energy in the Energy Policy Consultative Document;
RCEP—figures given by the Royal Commission on Environmental Pollution).
Values are million tons of coal equivalent.

	1985		2000	
	EPCD	RCEP	EPCD	RCEP
Primary energy (excluding non-energy)	380	360	490	390
Final demand	240	270	300	290

As the table shows, however, there is in 2000 a substantial difference between
the two estimates in terms of primary energy. This arises primarily from differing
views on the contributions to be expected before the end of the century from
renewable sources and combined heat and power, and from differences in conversion losses arising from a smaller electricity share of final demand (14%) compared
with EPCD (21%). Although the Royal Commission see a smaller electricity share
of final demand, they expect nuclear capacity in 2000 to increase substantially
over the present level to 27 GW in 2000—which is about one-third less than in the
reference case discussed in EPCD.

Beyond 2000

The Royal Commission's strategy for the period from 2000 to 2025 shows a continuing increase in final demand for energy but a considerable shift in the balance
between different energies. Oil and natural gas will be declining sharply, but major
increases are envisaged in the contribution of nuclear energy, coal in direct use and
as coal-based SNG, combined heat and power, all forms of renewable sources and
waste materials. The balance between final demand and primary energy required
implies lower conversion losses than might reasonably be expected, but the
elements of the strategy, with its reliance in 2025 on nuclear power (80 GW),
coal (210 million tons), coal-based SNG and a much enhanced contribution from
renewables, broadly match the elements of the strategy envisaged in this Green
Paper, even if the size of contribution from each source may be in doubt. In
particular, there is considerable doubt that the contribution from alternative
sources and from combined heat and power can reach the levels suggested by
the Commission.

The Commission's strategy of matching types of heat to market requirements
and of minimising losses through combined heat and power systems has much to
commend it. However, their assumption that combined heat and power schemes
and other forms of district heating will provide around 35% of the space and water
heating required in 2025 (25% from CHP, 10% from waste combustion/district
heating) represents a very ambitious target, particularly in view of the dislocation
involved in introducing the system into existing cities where the urban population
is already concentrated.

Little is at present known about the future cost of renewable sources and it is difficult to estimate how far they can contribute to supply in competition with other forms of energy, particularly nuclear.

The Royal Commission conclude that their analysis of a possible alternative strategy suggests that fast reactors might be avoided, at least to 2025, without recourse to imported oil. However, many of the factors underlying that alternative strategy appear to be optimistic. The growth of thermal reactor capacity in the world as a whole is likely to be very considerable. That growth will increase pressure on uranium supplies, which may be inadequate to support the demands made upon them. If only as an insurance policy, access to fast reactor technology may be important to ensure adequate supplies of energy into the twenty-first century. There is a considerable risk that an alternative strategy on the lines of the Royal Commission's might fail to achieve its targets. For all these reasons, it would not be wise to rely upon the alternative strategy to meet the UK's future energy needs.

Appendix 2

Alternative Ways of Calculating the Future Use of Energy in Coal Equivalent Terms

A Note by Peter Odell

TABLE A2.1 FUTURE ENERGY USE COMPARED WITH PRESENT SITUATION
Values are million metric tons (tonnes) of coal equivalent.

		Situation calculated for a future year		
		With development of nuclear power		Without nuclear
	Present situation			
	Case 1	Case 2*	Case 3†	Case 4
Electricity sector	30	130	63	55
Non-electricity sector	70	105	105	141
Total	100	235	168	196

*Nuclear power calculated on a notional fossil-fuel input base.
†Nuclear power calculated on heat value equivalence of electricity produced.

Three alternative ways of calculating future energy use compared with the present situation are compared in *Table A2.1*.

Case 1. This presents the situation in a contemporary non-nuclear economy. In this case, the primary fuels used for electricity generation account for 30% of total energy used.

Case 2. This shows the way in which official statistics of energy use in a future year (say 15–20 years hence) are presented. The non-electricity sector consisting of the direct use of fossil fuels (plus minor developments such as solar heat) grows modestly (from 70 to 150 mtce). The emphasis is on building up the electricity sector based on nuclear power. The value of the nuclear-based electricity is calculated in coal equivalent terms as though all the electricity were produced in conventional power stations working, on average, at a 33% rate of efficiency in converting fossil fuel to electricity. Overall, this case produces a 135% increase in energy use compared with the present situation.

Case 3. This is identical to *Case 2* except that the value of the nuclear electricity is now calculated on the equivalent heat value of the electricity produced. As this does not incorporate the value of the heat which would have been lost in the process of producing electricity, the total increase in energy use compared with the present is now only 68%—half as much as in *Case 2.*

Case 4. This shows what could happen if there were no nuclear electricity. In its absence, much less electricity will be produced because energy users, who in the nuclear electricity option would have used electricity to produce heat for space and water heating, etc., will now use fossil fuels for these purposes. For illustration purposes, here we assume that 25% of the additional amount of electricity produced in the nuclear-based system will be required. This is for uses for which there are no substitutes for electricity and the percentage taken is on the high side. This level of electricity production will require an input of 25 million tons of coal equivalent assuming a 33% average fuel conversion factor in the fossil fuel stations. Now, however, the economy will still be short of 25 mtce (= 33 − 8 mtce) of useful heat (compared with the nuclear option). To produce this heat in coal, oil or gas burning apparatus of average 70% efficiency will necessitate the use of about 36 mtce so that the non-electricity sector in this option will now need to grow from 70 to 141 mtce (= 105 mtce as in *Cases 2* and *3* plus this extra 36 mtce). The overall use of energy in this option is now 196 mtce: an increase of 96% on the current situation, but, nevertheless, well below the increase of 135% shown in *Case 2.*

In other words, the non-nuclear option *necessarily* means a lower rate of growth in the use of energy when compared with the way in which the statistics are now presented officially. The difference is large, so that the official presentations *severely overstate* the *apparent* future needs for fossil fuel in the event of the nuclear–electricity based economy not being developed. Indeed, the order of magnitude of the difference is so great that when it is seen in the context of the UK energy outlook, it completely eliminates the 'gap' between energy supply and energy demand which the Department of Energy considers will have to be filled by nuclear power by the 1990s. The significance of this certainly seems to justify the conclusion in my paper that the method of presenting the nuclear contribution is much more than a 'narrowly technical question'. On the contrary, one could argue that the way in which nuclear power's contribution to the energy economy is being presented in official statistics in itself serves to create the idea of an energy gap.

Appendix 3

The Breeding Principle and Fast Reactors*

THE BREEDING PRINCIPLE

Uranium-238, which plays little part in the fission process, can be converted by neutron capture and subsequent decay into plutonium-239 which, like uranium-235, is fissile. This happens in thermal reactors, and in fact early reactors such as those at Windscale were specifically designed to make plutonium-239 for weapons purposes. (The heat was not utilised.) In power reactors, the fuel is left in the reactor long enough for the plutonium produced to contribute substantially to the number of fissions, as indicated in *Figure A3.1*, but some plutonium nuclei remain in the fuel after irradiation. Plutonium-239 nuclei can capture neutrons without fissioning to become higher isotopes, only some of which are fissile, and part of the plutonium is in this form. The breeding of fissile plutonium partially counter-acts the loss of uranium-235 nuclei: in principle, the fuel could be reprocessed to remove the fission products, and restored to its original reactivity with less new uranium than would otherwise be needed. In fact, however, the repetition of this process would still allow only about 1-2% of the uranium nuclei in the fuel to be fissioned. The return is limited by the fact that in a thermal reactor the number of fissile nuclei created is less than the number of uranium-235 nuclei destroyed; that is, the ratio of these quantities, known as the 'breeding ratio' or 'conversion factor' is less than one.

If some way could be found to increase the breeding ratio above unity, it would mean that more fissile nuclei would be created than were destroyed, and it would offer the prospect of utilising the whole of the original uranium including the very large percentage of uranium-238 which cannot be fissioned in thermal reactors. This would be a great prize. The amount of uranium available in the world is large, but the element is usually found in poor-grade ores so that much effort has to be expended in winning it. The future demand for uranium is likely to be such that it may become both expensive and scarce. The ability to breed fissionable material from uranium-238 would transform the supply position, enabling the available supplies to be used perhaps 60 times as effectively in

*From the RCEP Sixth Report, chap. III, para. 108-117.

213

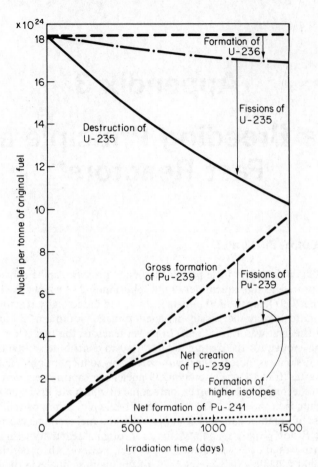

Figure A3.1 The changes in numbers of fissile nuclei in Magnox fuel during irradiation, showing the contribution made by plutonium-239.

producing energy. It is not surprising that major efforts are being devoted in the UK and in other countries to develop the means to achieve this.

In most thermal reactors, the neutron losses arising from metallic components in the core because of the use of a moderator are such that the breeding ratio is significantly less than unity. Two reactor types, the Candu and HTR, can be designed so that the breeding ratio is very close to unity, but this does not give the optimum economic performance and some fissile material is inevitably lost when the fuel is reprocessed, so that they are not self-sustaining in fuel. A breeding ratio sufficiently above unity for this to occur can, however, be obtained with an unmoderated reactor, where the nuclear chain reaction is sustained by fast neutrons alone. Such a reactor is called a 'fast reactor'. Its design poses difficulties because fission is much less likely to be induced by a fast neutron than by one moving at thermal speed. There must be a higher proportion of fissile

214

nuclei in the fuel and the neutron flux density (the number crossing a given area of core in unit time) must be much higher. In practice, this means that the fuel must contain a substantial proportion of plutonium admixed with the uranium (from which more plutonium is bred) and that the reactor core must be very compact, resulting in a very high thermal power density. A fast reactor that can breed enough plutonium to keep itself refuelled, and have some to spare, is called a 'fast breeder reactor' or FBR. We shall be mainly concerned with a particular development in which a liquid metal (normally sodium) is used as the coolant, although other systems are possible in principle, such as gas cooling.

Before describing the fast breeder reactor, we briefly mention an alternative approach to the higher utilisation of the nuclear fuel which involves the use of thorium as well as uranium in a thermal reactor. Thorium consists mainly of thorium-232, and like uranium-238, this isotope is not fissile but it is fertile, and can capture a neutron to form thorium-233, which decays by two successive β emissions to form uranium-233. This isotope of uranium does not occur in nature, but it is long-lived (half-life 159 000 yr) and fissile. A reactor with good neutron economy such as Candu or the HTR can be fuelled with a mixture of highly enriched uranium and thorium, and it can be designed to have a breeding ratio of nearly unity or even slightly above*, together with a high burn-up so that many of the uranium-233 fissions take place without the fuel having to be removed from the reactor and reprocessed. As a result of the utilisation of the thorium, the amount of uranium fed to a system of reactors burning both uranium and thorium would be less by a factor of about five than that needed with uranium-burning thermal reactors alone. But the technology for reprocessing and fabricating thorium fuel is hardly developed as yet, and although the thorium fuel cycle could make a substantial contribution in the longer term to the conservation of uranium, it cannot use the large stocks of uranium-238 which depend upon the plutonium-burning fast reactor for their conversion to useful energy.

THE LIQUID-METAL-COOLED FAST BREEDER REACTOR

The first nuclear reactor ever to produce electricity in the USA (albeit on a very small scale) was of this type, and in the UK the small Dounreay Fast Reactor went critical nearly two years before any of the CEGB's Magnox stations. But two serious accidents, involving partial fuel meltdowns, in US reactors delayed the programme there and construction work on a 350 MW prototype at Clinch River will not start before next year. The major development effort has been in Europe, which possesses little uranium ore, and thus has a greater need to use uranium economically than the USA. Medium-sized prototypes have been built in Britain, France and the Soviet Union, and one is under construction in West Germany. The Prototype Fast Reactor (PFR) at Dounreay (*Figure A3.2*) has a maximum rating of 250 MW, but it is envisaged that a full-scale commercial fast reactor (CFR) would be of about 1300 MW in order to drive two standard 660 MW turbo-alternators.

*Thermal fissions of uranium-233 produce on average more neutrons than those of uranium-235 or plutonium-239.

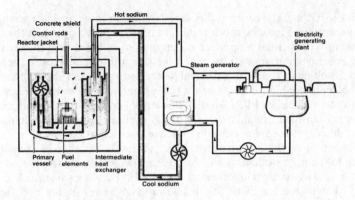

Figure A3.2 Schematic diagram of the Prototype Fast Reactor (United Kingdom Atomic Energy Authority).

The fuel in the centre of the core is a mixture of uranium and plutonium oxide, typically about one-fifth plutonium, or about five tonnes in the CFR design. This provides the majority of the reactivity. Surrounding this 'mixed oxide' central region are 'blanket' regions in which 'depleted' uranium, that is uranium from which the 235 isotope has been largely removed, absorbs neutrons and is converted to plutonium-239. It also acts as a neutron reflector. The fuel is removed at intervals, and the newly created plutonium is extracted in the reprocessing plant. As its name implies, the FBR breeds somewhat more plutonium (perhaps 10–20%) than is consumed in fission. Once a reactor has been built and provided with its initial inventory of plutonium (together with a margin of about 35% to allow for the amount needed in other parts of the fuel cycle), it needs no further supply of uranium-235 or plutonium, but will run indefinitely on the depleted uranium that is left over by the thermal reactors, and produce a net output of plutonium that can be used in other reactors. However, in order to provide enough plutonium to form the initial inventory for an FBR (about 4 tonnes per GW of electrical output), it is first necessary to run a thermal reactor of equivalent output for many years. This number depends on the reactor type: for a Magnox reactor or reactors it would be about 7 years, but for an AGR, 30 years, assuming a 66% capacity factor*. Thus although, in the long run, FBRs offer the prospect of near independence from uranium supplies, in the short term a large FBR programme will necessitate a big thermal reactor programme and hence a big uranium demand.

We have noted that the core of the reactor is necessarily very compact and highly rated in terms of heat output. The power density of the core of the PFR (which is comparable to that expected in the CFR) is about 390 kW/litre, compared with 11 kW/litre in the SGHWR core and only 1 kW/litre in Magnox reactors. (By way of comparison, the power density is about 0.8 kW/litre in a

*Capacity factor = (actual amount of electricity generated during a year) ÷ (the maximum possible amount from running at the nominal rating all the time).

modern coal-burning domestic fireplace.) As a result, it is difficult to arrange an adequate flow of either gas or water to remove the heat, and all the current fast reactors employ a liquid metal for this purpose. The high heat conductivity of the metal, usually liquid sodium, and the fact that the system is unpressurised are important advantages. For example, after reactor shutdown, natural convection alone will remove the fission product decay heat without the need for pumping. The use of sodium as coolant raises some problems. . . . One of these is that because sodium and water react chemically if they come into contact, and because the heat exchangers have high-pressure steam on one side and low-pressure sodium on the other, any slight leak might lead to a vigorous reaction and dangerous consequences for the core. Consequently, it is necessary to use a secondary sodium coolant circuit to convey heat from the pool of primary sodium in the core to the steam in the heat exchangers. Leaks have in fact occurred in these heat exchangers on the PFR at Dounreay and have delayed operation at high power levels by nearly two years.

Like the thermal reactor, the fast reactor is possible because of the accident of nature that some of the neutrons are delayed. However, if there were a net addition of reactivity so large and so fast that the normal control mechanisms could not deal with it, then there exists the theoretical possibility that there could be formed a nuclear assembly that was critical on prompt fast neutrons alone. This would lead to what is technically a nuclear explosion, though the growth of the chain reaction would be slow compared with that which occurs in a nuclear bomb and the energy release would be correspondingly less. It is not yet clear whether a nuclear explosion would vaporise fuel; it is currently assumed that this could occur and the reactor containment is designed to cope with such a contingency. However, if it failed to do so, then not only iodine and caesium, but substantial quantities of non-volatile fission products such as strontium, as well as plutonium, would be released. If the reactor were in a populous area, the number of casualties could be very great. The reason why this can occur in a fast reactor, but not in a thermal one, is that in the former the fuel is not initially in its most reactive state. If all the fuel in a thermal reactor were to melt into one mass, it would be less reactive because there would be no moderator to enhance criticality. But if all the fuel in a fast reactor were to melt into a single compact mass, it would be very much more reactive.

The two partial meltdowns in FBRs in the USA were contained, and there was no release of radioactivity. But an uncontained meltdown could be so serious in its consequences that it is generally accepted that fast reactors cannot be major contributors to a power programme until the processes underlying the change of geometry are well understood. There is an extensive research programme in the field, but it is not yet clear whether it will prove possible so to design fast reactors as to rule out in principle the possibility of a sudden increase in nuclear power that would be so great as to rupture any feasible containment.

The fast breeder depends on plutonium for its fuel. This raises a number of problems, some of which we consider in detail elsewhere. These include its radiotoxicity and the consequent need for elaborate precautions in handling it; and the risk that it may be used for nefarious purposes, which requires that it should be carefully safeguarded. It should also be noted that plutonium is essentially a man-made element and that its supply depends on the operation of

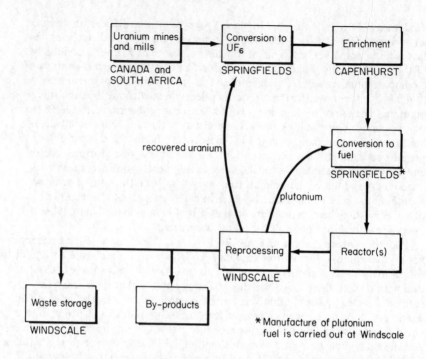

Figure A3.3   The nuclear fuel cycle (redrawn from Patterson, W. C., *Nuclear Power,* Penguin, 1976).

reprocessing plants. If such a plant was put out of action for a long time, as it could be, for example, by an accident leading to substantial contamination, a serious supply situation might well be created unless large stocks of plutonium existed to cover the contingency or there were alternative sources of supply. The nuclear fuel cycle is shown in *Figure A3.3.*

Appendix 4 Energy Policy Matrix

Inputs		Economic projections	Political projections	Policy decisions (medium term to year 2000)
Energy supply	to 3% per annum	Rising real costs; increased per cent GNP	Rising living standards assumes greater energy inputs	Raise energy output to 490 mtce despite higher costs
Electricity supply	to 3% per annum	High opportunity costs	Increased proportion of energy supply	Replace direct use of fossil fuels
Nuclear programmes (a) thermal (b) FBR	1 × 1250 MW every 2 years	Thermal preferred until U_3O_8 reaches \$80.16	Fast breeder preferred to sustain nuclear option	5–12 GW FBR decision after public inquiry
Energy efficiencies	Possible improvements projected	Large benefits possible (if electrical end use constrained)	No major reorganisation of energy supply	No overall rise in energy efficiency
Fuel mix	Indicative; targeting	Nuclear plus coal for high case; nuclear versus coal for low case	Nuclear preferred to coal by CEGB and Dept of Energy	Coal (170 mtce), oil (150 mtce), nuclear (95 mtce), Natural gas (50 mtce)
Security of supply		Nuclear replaces oil in the longer term	FBR programme essential to avoid uranium supply constraint	Basis of EPCD priorities
Prices	Reflect resource costs	Average real rise of 2½% per annum	Market prices determine energy supplies; minimum use of price regulator	Uncertainties in future prices will continue
Costs	Accounting cost approx; real costs	World oil price as marginal cost	Low costs preferred and subsidy for nuclear power	Dept of Energy assumes rise in real costs of 50% by the year 2000

Inputs		Economic projections	Political projections	Policy decisions (medium term to year 2000)
Instruments of policy	Market forces predominantly	Forecasting outputs	Investment control	Long-term investment in nuclear power
R & D	Half to nuclear	Opportunity costs of nuclear R & D rising	No significant re-allocation to renewable energy sources	FBR expenditures to be maintained (expand?)
Capital provision	Remains high	Assumed rise but depends on GNP growth; capital scarcity may arise?	Main instrument of central control	No forecasts, but will rise in state sector
Conservation	Rising prices main instrument	Limited change before costs to supply industries felt	Voluntary programme only; unlikely to realise 23% target	23% cut in heat supplied
Reprocessing of nuclear wastes		High costs with losses in domestic contracts	Essential for nuclear option	Expand output at Windscale to 600 tonnes per annum
Nuclear waste disposal		Rising costs; costs of uncertainty	No delay in FBR programme	Continued research and search for solution
Radioactive pollution		Rising costs	Present measure for control adequate	No change
Nuclear safety		Rising costs	UK reactors are safe; risks small	No change
Decommissioning of nuclear stations		Costs unknown, but could be large	Not yet an issue	
Nuclear proliferation		Balance of payments gains from foreign contracts in nuclear materials	Rising dangers not a cause for policy changes	International agreements, not unilateral action
Environmental effects		Social costs to the consumer	Cautious approach; legislation and action lag behind requested steps	

	Resource allocation by prices	Security of supply	Govt. initiated and intervention	Nuclear power and ES institution	Environmental and social factors
Civil liberties } Terrorism	Costs to nuclear programme will be significant		Pressure to halt or abandon nuclear programme not accepted		No information given
Consumer interest	Higher energy costs		Increased concern and participation partially recognised		Principle of public inquiries established in areas of special concern
Nuclear establishment	Redeployment could bring benefits		National needs for nuclear option take priority		Political requirements influence/determine decision making
Industrial capacity for nuclear power	Weakened by lack of orders		Requires government approval of more nuclear studies		

PRIORITIES IN POLICY FORMULATION

	Resource allocation by prices	Security of supply	Govt. initiated and intervention	Nuclear power and ES institution	Environmental and social factors
1959–65	√√√	√	√	√√	
1967–75	√√√	√	√√	√√√	√
1975–80	√√	√√√	√√	√√√	√
1980–2000	√√	√√√	√√	√√√	√

Appendix 5

A Potential Fast Breeder Accident at Kalkar. A Summary of a Recent PERG Study

This report contains a summary of the results of a study[1] recently completed by PERG, together with some observations on those results. The study was motivated by the attendance of PERG members at the public hearings organised in Brussels by Commissioner Brunner[2]. Consequences of a postulated FBR accident were determined using a computer program[3] written by the UK Atomic Energy Authority and data from a recent report[4] of the UK National Radiological Protection Board.

THE POSTULATED REACTOR ACCIDENT

We have considered an accident drawn from the upper limit of severity of the accidents considered by the NRPB and specified for them by the UK Nuclear Installations Inspectorate. We have assumed a sodium-cooled FBR of 1300 MW electrical output. This reactor was assumed to suffer an accident such that 10% of the core became vaporised, the remaining 90% of the core being molten. Total failure of containment was assumed, this resulting in release of a cloud of radioactivity to the atmosphere. A computer program then predicts the consequences to the public in terms of casualties and long-term health effects.

DISPERSION OF THE RADIOACTIVE CLOUD

After the cloud has left the reactor, it will travel downwind and gradually become dispersed. *Figure A5.1* shows diagrammatically how this process occurs and also how people become irradiated.

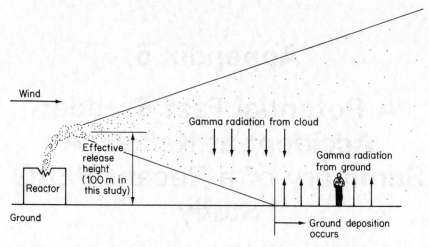

Wind

Gamma radiation from cloud

Effective release height (100 m in this study)

Gamma radiation from ground

Reactor

Ground

Ground deposition occurs

Figure A5.1 Dispersion of the radioactive cloud. Note that, in addition to the gamma doses shown, individuals will suffer a substantial dose from the inhalation of radioactive particles.

The degree to which the cloud becomes spread out is determined by the state of turbulence of the atmosphere. This is calculated by use of the computer model. We present here the results for two cases:

Weather category 4 (wind speed 6 m s^{-1}): This occurs typically 50–60% of the time and represents an intermediate rate of cloud spreading.

Weather category 6: This occurs typically 5–10% of the time and represents a low rate of cloud spreading. High concentrations of radioactivity may be carried considerable distances. This type of weather tends to persist for a few hours only.

CONSEQUENCES TO THE PUBLIC

Three categories were considered:

Early deaths: These are deaths beginning in the first few days or weeks and extending over the first year. The bone marrow and lung are the most affected organs.

Lung morbidity: This refers to respiratory impairment leading to a reduction in the quality of life and/or life shortening.

Cancer deaths: Cancers begin to appear after a latent period of 5–15 years. The most likely cancers are bone cancer, liver cancer, lung cancer and leukaemia, although other types will occur. These cancers can be considered as reducing

224

life expectancy and affect younger people most in this respect. A 1 year-old person can expect perhaps a 13 year reduction in life expectancy where a 50 year-old person would expect a 1 year reduction in the same circumstances.

In addition, there will be illnesses from which people recover and hereditary effects. These we will estimate at a later date as resources permit.

Number of People Suffering Death or Damage to Health

The population which is affected will be that living in the sector into which the cloud is carried. Consequences to the public were determined up to a distance of 150 km, on the assumption that complete evacuation of contaminated areas occurred after 1 day. We present here results for sector 5, which is the sector yielding the greatest number of health effects.

Health effects for wind into sector 5	Weather category	
	4	5
(a) Early deaths	450	125 000
(b) Cases of lung morbidity	1 840	389 000
(c) Deaths from cancer	81 700	467 000
(d) Persons suffering lung morbidity and/or cancer death	83 160	810 650
(e) Total of health effects [(a) + (d)]	83 610	935 650

CONTAMINATION OF THE GROUND

Radioactivity which is deposited on the ground will emit dangerous gamma radiation and severely contaminated areas must be evacuated. This deposited radioactivity will be removed relatively slowly by natural processes. We have computed the area of land within 100 km of Kalkar, in the affected sector, which must be abandoned for at least 20 years. This area is as follows:

Weather category 4: 334 km^2
Weather category 6: 227 km^2

The contaminated area will extend considerably beyond 100 km, for instance at a distance of 130 km a strip of land must be evacuated for at least 20 years which is 11 km and 8 km in width for weather categories 4 and 6 respectively. Contamination may extend for hundreds of kilometres, but the computer model used becomes unreliable for such distances and thus we offer no estimates.

COMPARISON OF PERG RESULTS WITH THOSE OF NRPB

The NRPB study is a valuable contribution to public knowledge on FBR safety and its assumptions are stated clearly in the summary report.

Our findings, using the same computer program, nevertheless include numbers of health effects substantially in excess of those listed by NRPB.

There are a number of differences between our study and the NRPB, for example our use of a Gaussian lateral distribution for cloud spreading (NRPB assume even distribution across a 30 degree sector), and we use an RBE of 20 for alpha radiation (a measure of its radiotoxicity), where the NRPB use the value of 10 now considered too low by ICRP (International Commission on Radiological Protection).

However, it is the inclusion of category 6 weather conditions that gives rise to the big increase in effects—the NRPB do not include figures for this category.

IMPLICATIONS OF OUR RESULTS

The international implications of an accident at Kalkar are evident. The prototype FBR under construction at Kalkar involves a collaboration of Belgium, Germany and the Netherlands. Such an accident would leave in its wake huge social and economic costs. If resources permit, we will investigate these costs at a later date.

The enormity of this accident raises an important question. It is generally accepted that probabilities estimated beyond the one in a million mark make little sense where complex technological systems are concerned. The nuclear industries and controlling authorities have argued that any accident risk is acceptable provided the probability is sufficiently low. This risk is defined as probability × consequences.

Where accidents of this enormity of consequence are concerned the question arises: is *any* such risk justified when the probability cannot be adequately defined? The ruling of the Freiburg High Court, on 15 March 1977, for the Wyhl case (a PWR reactor) was that the 'remnant of risk', however small, was not justified in view of the magnitude of the consequences of a severe accident[5]. It may be that other courts will take this view.

REFERENCES

1. *A Study of the Consequences to the Public of a Severe Accident at a Commercial Fast Reactor Located at Kalkar, West Germany*, PERG RR-1, Report of Political Ecology Research Group, Oxford, January 1978
2. Open Discussions on Nuclear Energy, organised by the Commission of the European Communities, chaired by Dr Guido Brunner, Commissioner for Energy, Research, Science and Education.
   First session: 29 Nov-1 Dec 1977 on energy needs and supplies for the rest of the century—the role of nuclear energy.
   Second session: 24–26 Jan 1978 on economic growth and energy options—implications for safety, health and environmental protection

3. Computer program TIRION, written by the Safety and Reliability Directorate of the UK Atomic Energy Authority as a tool for the estimation of the radiological consequences of an atmospheric release of radioactivity. For further information consult reports SRD R 62 (Nov 1976) and SRD R 63 (Oct 1976) written by Dr G. D. Kaiser of SRD.
4. *An Estimate of the Radiological Consequences of Notional Accidental Releases of Radioactivity from a Fast Breeder Reactor*, Report NRPB R-53 of the UK National Radiological Protection Board, August 1977
5. Ruling of the Freiburg High Court on the siting of the Wyhl reactor, March 1977 (Verwaltungsgericht Freiburg—see Whyl Urteil, Dreisam Verlag, Shwaighofstrasse 6, 78 Freiburg, West Germany)

Index

230

231

Torness, cost of   133
Traube, K.   83

UKAEA, *see* AEA
Union of Concerned Scientists   82
UNSCEAR   112-4
Uranium
   U-233   184
   U-235   179
   U-238   184
   Doppler coefficients   63
   Institute   101
   miners   109-10
   price and scarcity   163
   waste   148-9

WASH Reports 740 and 1420   82
Waste, *see* Nuclear
Weinberg, A.   4
Wilson, R.   62, 69-73
Windscale   193, 196
Windscale Public Inquiry   1, 76, 79,
   171
   access to official information   83-4
   BNFL and AEA evidence   169
Workshop on Alternative Energy
   Strategies   53
World Energy Conference (conservation
   studies)   17

X-rays   103

232